Environmental Resilience and Food Law

Agrobiodiversity and Agroecology

Advances in Agroecology

Series Editors

Clive A. Edwards
The Ohio State University, Columbus, Ohio

Stephen R. Gliessman
University of California, Santa Cruz, California

Sustainable Agriculture and New Biotechnologies
edited by Noureddine Benkeblia

Global Economic and Environmental Aspects of Biofuels
edited by David Pimentel

Microbial Ecology in Sustainable Agroecosystems
edited by Tanya Cheeke, David C. Coleman, and Diana H. Wall

Land Use Intensification
Effects on Agriculture, Biodiversity, and Ecological Processes
edited by David Lindenmayer, Saul Cunningham, and Andrew Young

Agroecology, Ecosystems, and Sustainability
edited by Noureddine Benkeblia

Agroecology
A Transdisciplinary, Participatory and Action-oriented Approach
edited by V. Ernesto Méndez, Christopher M. Bacon, Roseann Cohen, Stephen R. Gliessman

Energy in Agroecosystems
A Tool for Assessing Sustainability
by Gloria I. Guzmán Casado and Manuel González de Molina

Agroecology in China
Science, Practice, and Sustainable Management
edited by Luo Shiming and Stephen R. Gliessman

Climate Change and Crop Production
Foundations for Agroecosystem Resilience
edited by Noureddine Benkeblia

Environmental Resilience and Food Law
Agrobiodiversity and Agroecology
edited by Gabriela Steier, Alberto Giulio Cianci

Political Agroecology
Advancing the Transition to Sustainable Food Systems
by Manuel González de Molina, Paulo F. Petersen, Francisco Garrido Peña, Francisco R. Caporal

For more information about this series, please visit:
https://www.crcpress.com/Advances-in-Agroecology/book-series/CRCADVAGROECO

Environmental Resilience and Food Law

Agrobiodiversity and Agroecology

Edited by

Gabriela Steier

Alberto Giulio Cianci

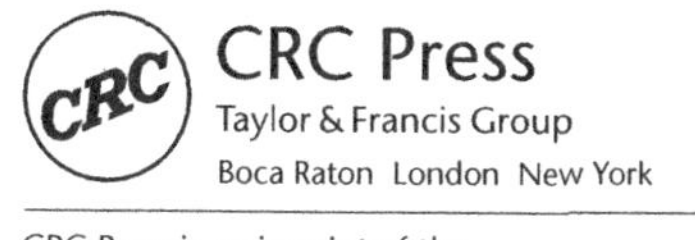

CRC Press is an imprint of the
Taylor & Francis Group, an **informa** business

CRC Press
Taylor & Francis Group
6000 Broken Sound Parkway NW, Suite 300
Boca Raton, FL 33487-2742

First issued in paperback 2022

ISBN-13: 978-1-138-33606-3 (hbk)
ISBN-13: 978-1-03-233807-1 (pbk)
DOI: 10.1201/9780429443350

Publisher's Note

The publisher has gone to great lengths to ensure the quality of this reprint but points out that some imperfections in the original copies may be apparent.

Library of Congress Cataloging-in-Publication Data

Names: Steier, Gabriela, editor. | Cianci, Alberto Giulio, editor.
Title: Environmental resilience and food law : agrobiodiversity and agroecology / edited by Gabriela Steier, Alberto Giulio Cianci.
Description: Boca Raton : Taylor & Francis, 2020. | Series: Advances in agroecology | Includes bibliographical references and index.
Identifiers: LCCN 2019017168| ISBN 9781138336063 (hardback : alk. paper) | ISBN 9780429443350 (e-book)
Subjects: LCSH: Food law and legislation. | Agrobiodiversity conservation--Law and legislation. | Sustainable agriculture--Law and legislation.
Classification: LCC K3626 .E58 2020 | DDC 343.07/6--dc23
LC record available at https://lccn.loc.gov/2019017168

Visit the Taylor & Francis Web site at
http://www.taylorandfrancis.com

and the CRC Press Web site at
http://www.crcpress.com

Dedication

This book is dedicated to Regina, Liviu, Morrice and Michael Fany, Mischu, Adela and Lotzi, with love

Gabriela Steier

For Viktoryia, wishing that she achieves all her dreams, like new leaves on a tall, strong birch

Alberto Giulio Cianci

Contents

Preface

> How sad to think that nature speaks and mankind doesn't listen.
>
> **Victor Hugo, 1820**

Agrobiodiversity and agroecology go hand in hand with the goals of promoting environmental resilience in international food systems and climate-change-resilient food policy. This book seeks to contextualize how various legal frameworks address agrobiodiversity and agroecology for students, practitioners, and scholars. Some of the chapters focus on the legal regulation of agroecology from a food law perspective, but other chapters are geared toward providing regulators, lawmakers, and lawyers with the scientific and policy background of those concepts. The intended audience encompasses all readers of the LITES series, including students, legal scholars, non-legal academics, advocates for food-system resilience, agroecology and environmental conservation, practitioners in any of the cross-disciplinary areas relating to food policy, and all those who seek to deepen their understanding of the concepts and trends surrounding agroecology and agrobiodiversity.

Environmental Resilience and Food Law: Agrobiodiversity and Agroecology is a collection of essays with editors' notes tracing the global trends of environmental resilience and food law in agrobiodiversity and agroecology. Breaking down these charged terms leaves us with an analysis of the challenges of implementing agrobiodiversity and agroecology through environmental conservation and through food law. How do these terms fit together?

Each chapter in this book answers different questions, all of which contribute to the overall lessons about how environmental resilience and food law come together through agrobiodiversity and agroecology. The following essential questions are developed further in each of these chapters, corresponding with chapter numbers:

1. What do we know about promoting environmental resilience through agroecology and biodiversity?
2. How does agriculture law address agroecology and agrobiodiversity? Where does it overlap with private law?

3. Can a rights-based approach further the UN's Sustainable Development Goals and use existing laws to protect agroecology and agrobiodiversity?
4. How can resources from food production and food waste and food loss be saved to achieve more sustainability?
5. What are the impacts of indigenous peoples on biodiversity conservation efforts in Africa and how are they addressed?
6. How can persons with disabilities experience more inclusive participation in aspects of agroecology and agrobiodiversity?
7. How do Mediterranean olive harvests illustrate the rich history and modern challenges of agroecology and agrobiodiversity?
8. What are the most important Asian trends in regulation reform in ensuring these values?
9. How can science (and the regulation thereof) prevent the losses of crop diversity?

As editors, we are humbled by the expertise of the contributing authors and honoured to have had the pleasure of assembling their scholarship in this volume. We invite our readers to ask probing questions, to engage with the material, and to challenge the solutions outlined therein. The only way to do the contributing authors justice is by reading their chapters carefully. Nonetheless, through a proactive and inquisitive approach to the challenges that increased regulation of agrobiodiversity protection and agroecology establishment provoke, we hope to contribute to the development of more sustainable and resilient food systems around the globe.

Gabriela Steier and Alberto Giulio Cianci, Boston 2019

Acknowledgments

The editors thank Alice Oven from CRC Press at Taylor & Francis for her unwavering support and Prof. Steven Gliessman for his valuable advice. All of the authors deserve added gratitude for their contributions to this book. The editors also thank their interns, Alex Cherry and Taylor Corn, who helped tremendously in editing of the received chapters.

Special thanks go to Regina and Liviu, Morrice and Michael for creating time for Gabriela to dedicate to this book, for their support and encouragement. An additional word of gratitude goes to Barbara and Murray Miller.

Another mention of gratitude goes to the editors' students, who, during joint webinars organized for them, expressed their views about many issues related to the content of this book.

About the editors

Dr. iur. Gabriela Steier, LL.M., Esq. is a lawyer, scholar, and educator focusing on food law and policy in the USA and the EU. She has published various books and articles about international food regulation, farm animal welfare, GMOs, food integrity, and sustainable agriculture. She is a part-time Professor at Northeastern University in Boston, an Adjunct Professor at the Duquesne University School of Law in Pittsburgh, and was Visiting Professor at the University of Perugia in Italy. She holds an LL.M. in Food and Agriculture Law from the Vermont Law School, a Doctorate in Comparative Law from the University of Cologne, a Juris Doctor from the Duquesne University School of Law, and a Bachelor of Arts from Tufts University. Gabriela lives in Boston and can be contacted at g.steier@foodlawinternational.com.

Avv. Prof. Alberto Giulio Cianci is an Associate Professor of Private Law Institutions and Advertising and Consumer Protection in the University of Perugia, Italy. He has published four books: on private law and personal freedom (*Diritto privato e libertà costituzionali*, vol. 1, *Libertà personale*, Naples, 2016), on advertising and consumer contracts (*Comunicazione pubblicitaria e contratti del consumatore*, Milan, 2013), on reciprocity and its current dimension in private law (*Soggetto straniero e attività negoziale*, Milan, 2007), and on law and domestic violence (*Gli ordini di protezione familiare*, Milan, 2005), as well as many essays on other topics of private law. In 2016, he qualified as Full Professor of Private Law. He practices as an attorney in Rome and can be contacted at alberto.cianci@unipg.it.

Contributors

Rosemary E. Agbor
College of Education
Grambling State University
Grambling, Louisiana

Michael Blakeney
School of Law
University of Western Australia
Crawley, Western Australia

Susannah Chapman
TC Beirne School of Law
University of Queensland
St Lucia, Queensland

Alexander Cherry
Peace Corps
Porto-Novo, Benin

Wele Elangwe
School of Graduate Studies
University of Maryland Eastern Shore
Princess Anne, Maryland

Paul J. Heald
College of Law
University of Illinois
Champaign, Illinois

Ivan K. Mugabi
School of Law
St. Augustine International University
Kampala, Uganda

Sumit Saurav
School of Law
KIIT University
Bhubaneswar, India

List of abbreviations

AFNs	Alternative Food Networks
CBD	Convention on Biological Diversity
DUI	Doing, Using, and Interacting Model
EC	European Commission
EU	European Union
FAO	Food and Agriculture Organization of the United Nations
FLW	Food Loss and Waste
GIs	Geographical Indications
IGC	Intergovernmental Committee on Intellectual Property and Genetic Resources, Traditional Knowledge and Folklore
ILO	International Labour Organization
IP	Intellectual Property
IUCN	International Union for the Conservation of Nature
LAK	Local Agricultural Knowledge
PDO	Protected Designation of Origin
PGI	Protected Geographical Indication
STI	Science, Technology, and Innovation Model
TK	Traditional Knowledge
TRIPS	Agreement on Trade-Related Aspects of Intellectual Property Rights
UN	United Nations
WIPO	World Intellectual Property Organization
WTO	World Trade Organization

part one

Fundamentals of legal protection of agrobiodiversity and agroecology

chapter one

Local agricultural knowledge and food security

Michael Blakeney

Contents

1.1 Introduction

This chapter provides a literature review of local agricultural knowledge in the context of food security. The Food and Agriculture Organization of the United Nations (FAO) estimates that about 795 million people were chronically undernourished in 2012–2014.[1] In 1996, the World Food Summit defined food security as "when all people at all times have access to sufficient, safe, nutritious food to maintain a healthy and active life."[2] With 70 percent of the world's extreme poor and food-insecure populations living in rural areas, the role of agriculture, which is the predominant economic activity in those areas, is crucial for the eradication of poverty and food insecurity.

Smallholder farmers increasingly cultivate marginal lands which are particularly vulnerable to climate change.[3] The FAO has observed that climate change will likely cause "many of today's poorest developing countries ... to be negatively affected in the next 50–100 years, with a reduction in the extent and potential productivity of cropland."[4] A 1996 FAO study estimated that the largest reductions in cereal production, averaging around 10 percent, will occur in developing countries.[5] To put this in perspective, a projected 2 percent to 3 percent reduction in African cereal production for 2020 has been estimated to be enough to put 10 million people at risk. Among the particularly vulnerable are low- to medium-income groups living in flood-prone areas who may lose stored food or assets; farmers who may have their land damaged or submerged by rising sea-levels; and fishers who may lose their catch to shifting water currents or to flooded spawning areas.

Compounding these problems are the estimates, according to Pinstrup-Anderson et al., that, at the current rate of global population increase, "the global demand for cereals will increase by 20 percent between 1995 and 2020, [and so] net cereal imports by developing countries will have to double to meet the gap between production and demand."[6] Currently, the developing world is a net importer of 88 million tons of cereals a year at a cost of US$14.5 billion, and its demand for cereals will increase by 40 percent between 1995 and 2020.[7]

The conventional policy approach to guarantee food security is to promote technological improvements in agriculture. The massive increases in food productivity in the 30 years between 1960 and 1990, often described as the "Green Revolution," were achieved by the development of high-yielding crop varieties, as well as massive increases in fertilizer and insecticide use. By 1990, however, it had become apparent that reliance upon the chemically nurtured, high-yielding crop varieties that had precipitated the Green Revolution was no longer economically or environmentally acceptable.[8] Therefore, in order to meet the food security needs of the next 30 years and create wealth in poor communities there would need to be increases in agricultural productivity on the presently available land without compromising the natural resource base.[9]

Governments have also introduced hybridized crop varieties, often developed by multinational life-science corporations, but these were often vulnerable to pest infestation or disease.[10] In response, the local knowledge and agricultural practices of traditional farming communities have begun to be looked at as more serious assets for establishing and maintaining sustainable agricultural systems.[11] As this chapter will indicate, an essential component of food security is the contribution that traditional farmers have made in identifying and conserving useful biological material, embodied in biotechnological innovations. Implementing agricultural advances depends upon the existence of appropriate legal instruments, the recognition and enablement of change, the local knowledge of farmers, as well as their understandings and adaptations of scientific knowledge.[12] This chapter reviews the literature surrounding local knowledge and its contribution to agricultural innovation and food security.

1.2 *Local knowledge, traditional knowledge, and intellectual property*

When evaluating the contribution that local knowledge can make to agricultural innovation and food security, a distinction has to be drawn between the terms "local knowledge" and "indigenous knowledge" or "traditional knowledge." It is also useful to consider the extent to which these knowledges might be classified as "intellectual property."

Confusion between these terms has often arisen during the negotiations within the Intergovernmental Committee on Intellectual Property and Genetic Resources, Traditional Knowledge and Folklore (IGC) of the World Intellectual Property Organization (WIPO), which is seeking to negotiate three treaties that would recognize, regulate, and protect traditional knowledge, traditional cultural expressions, and genetic resources.[13] To assist the IGC in its task of formulating draft texts for these treaties, the WIPO Secretariat has prepared a Glossary of Key Terms Related to Intellectual Property and Genetic Resources, Traditional Knowledge and Traditional Cultural Expressions.[14] In its Glossary, the WIPO Secretariat reiterates that "there is as of yet no accepted definition of traditional knowledge (TK) at the international level."[15] It draws a distinction between TK "as a broad description of subject matter" as that which:

> generally includes the intellectual and intangible cultural heritage, practices and knowledge systems of traditional communities, including indigenous and local communities (traditional knowledge in a general sense or *lato sensu*). In other words, traditional knowledge in a general sense embraces the content of knowledge itself as well as traditional cultural expressions, including distinctive signs and symbols associated with traditional knowledge.

As opposed to TK in "international debate," where it, "in the narrow sense," refers to:

> knowledge as such, in particular the knowledge resulting from intellectual activity in a traditional context, and includes know-how, practices, skills,

> and innovations. Traditional knowledge can be found in a wide variety of contexts, including: agricultural knowledge; scientific knowledge; technical knowledge; ecological knowledge; medicinal knowledge, including related medicines and remedies; and biodiversity-related knowledge, etc.[16]

A number of problems have been identified with the concepts of "indigenous" and "traditional" knowledge. One political obstacle to their protection is that the terms have become associated with the right of peoples to self-determination. In 1993, Erica-Irene Daes, Special Rapporteur of the UN Sub-Commission on the Prevention of Discrimination and Protection of Minorities and Chairperson of the Working Group on Indigenous Populations, observed in a report that "the protection of cultural and intellectual property is connected fundamentally with the realization of the territorial rights and self-determination of indigenous peoples."[17] These principles are explicitly reaffirmed in Article 31 of the UN's Declaration on the Rights of Indigenous Peoples from 2007, which recognizes the rights of indigenous peoples to "maintain, control, protect and develop as intellectual property their cultural heritage, traditional knowledge, and traditional cultural expressions."[18]

Daes's concept of "indigenous" is grounded upon the idea of a distinctive culture, based on long-held traditions and knowledge, that is essentially connected to a specific territory. Le Gall, on the other hand, points out that confining the protection of traditional knowledge to indigenous creations overlooks the contributions of the creations and knowledge of native populations which might have arrived in countries as slaves, refugees, or indentured workers.[19]

Furthermore, according to Dove, it is problematic to conflate "local" or "traditional" knowledge with "indigenous knowledge," as this term has a different meaning in developing Asian countries compared with settled colonies.[20] The governments of several Asian countries have expressed reservations about the applicability of the term "indigenous people," which in their views is more appropriate to use in the context of Euro-American colonialism.[21]

1.3 Legal transplants

In attempting to fit local, traditional, and indigenous knowledge into international legal instruments by establishing legal classifications, Antons points out, communities are expected to live up to the expectations of outsiders, particularly lawyers and policy-makers, regarding the "authenticity" of their "traditional lifestyles."[22] Sillitoe has also suggested that it may be problematic to term "traditional" what is essentially an "admixture of local folk knowledge and extra-local scientific knowledge."[23] Another term for this amalgamation, according to Frossard, is "peasant science."[24] Dove has also suggested that the term "traditional" does not do justice to composite systems of agriculture,[25] which combine subsistence-oriented swiddens with market-oriented cash cropping.[26] Furthermore, the term "indigenous knowledge" fails to capture the historical movement and exchange of plant genetic resources between different parts of the world that resulted in today's local farmer knowledge and practices.[27]

The IGC indicated as early as its third session a proclivity to "leave specific determinations of the boundaries of protectable subject matter up to domestic authorities ... [allowing] terminology at the international level to be used more to express a common policy direction."[28] Therefore, a generalized definition of IP protection, especially at the international level, can be distinguished from the more precise definitions that are developed and applied on a case-by-case basis at the national or regional levels.[29]

Attempts to fit traditional, indigenous, and local knowledge into the framework of intellectual property, however, encounter the more fundamental problem that Drahos labels the "tragedy of commodification," where, according to the western tradition, the object of protection is separated from the environment in which it evolved.[30] It is feared that the WIPO IGC's attempts to fit these categories of knowledge within treaties that provide for their exploitation might promote a colonialist style of transmission of law codes from industrialized to developing countries,[31] and also interfere in intra-community disputes over the distribution of those legal benefits.[32] Thus Bragdon notes that the focus of discussions in the IGC concerning crop improvement by farmers focuses upon genetic resources per se rather than on the innovative process where all germplasm, traditional or modern, "is treated as a potential input for direct use or further improvement."[33]

Another consequence of the fitting of traditional knowledge into the international treaty-making process is that the customary law systems of traditional communities are invariably removed as being "inappropriate"[34] or "uncertain."[35]

1.4 *Local knowledge, innovation, and the informal economy*

Since 2001, the WIPO IGC has spent 34 sessions attempting to negotiate treaties dealing with traditional knowledge, traditional cultural expressions, and genetic resources, without any prospect of satisfactory conclusion, at least in the near future.[36] In part, this long-and-drawn-out negotiation can be attributed to the political challenges that some WIPO Member States have in recognizing the rights or existence of indigenous peoples. In part, the complexity of these negotiations can be traced to the difficulty of fitting traditional knowledge into the system of intellectual property, which largely follows the science, technology, and innovation (STI) model, which relies on incentivized innovation based in research and development in order to produce explicit and codified scientific and technical knowledge. A contrasting model, based on doing, using, and interacting (DUI), instead relies upon an experience-focused mode of innovation, which involves learning from informal interactions and often results in innovations with more tacit elements.[37] Of course, innovations to be protected as a category of intellectual property can contain elements of both models, but agricultural innovations of local communities are on the global scale almost entirely based on the DUI model and thus do not easily fit within the current intellectual property categories.

Although the informal economy, including agriculture, is estimated to exceed 90 percent of the economies of some developing countries, definitions of the informal economy remain elusive.[38] The concept of "informality" has been dated back to studies of African economies from the early 1970s.[39] An International Labour Organization (ILO) report in 1972 defined informal economic activities as those characterized by:

(a) ease of entry
(b) reliance on indigenous resources
(c) family ownership of enterprises
(d) small scales of operation
(e) labour-intensive and adapted technology
(f) skills acquired outside the formal school system
(g) unregulated and competitive markets[40]

These characteristics particularly typify agricultural activities in developing and least-developed countries.

Another feature of the informal economy is how it relates to intellectual property. In the formal sector, the intellectual property system is used to appropriate the products of technological or creative innovation through patents, trademarks, industrial designs, and copyright schemes.[41] In the informal sector, on the other hand, actors give little consideration to appropriating the products of innovation and use semi-formal or informal means of appropriation.[42] In the informal sector, a premium is placed on trust, personal relationships, social beliefs, values, and norms, and there is a greater absence of written agreements.[43] In any case, the actors in the informal sector are typically locked out of access to formal intellectual property rights, in part because they are unaware of the intellectual property system. Even when they are aware, though, they often lack the necessary legal skills and support organizations to secure formal intellectual property rights.

Producers' access to formal agricultural innovations in the informal economy is often facilitated by communications of agricultural extension officers[44] and field schools arranged by university agronomists.[45]

1.4.1 Utility of local knowledge

Agriculture in the developing world is largely dominated by small-scale farmers working in marginal environments using locally developed agricultural methods. These methods have evolved on their own over time and represent the experiences of farmers interacting with their environments to meet their subsistence needs.[46] For example, the strategies used to manage the toxicity of cassava in disparate marginal environments around the world depend upon the local agronomic knowledge in those environments.[47] Altieri writes that local agricultural knowledge develops from a combination of observation, experimentation with local seed varieties, and the "testing of new cultivation methods to overcome constraints such as soil infertility and pest infestations."[48] Such traditional crop-management practices have been identified as a rich resource for understanding the interactions between biodiversity and ecosystem function[49], which is crucial for the identification of principles needed to develop more sustainable agricultural systems.[50]

Altieri[51] pointed out that most traditional agroecosystems share a number of structural and functional similarities:

- high species numbers
- high structural diversity in time and space
- exploitation of the full range of local microenvironments
- maintenance of closed cycles of materials and waste through effective recycling practices
- complex biological interdependencies, resulting in a high degree of natural pest suppression
- dependence on local resources and human and animal energy, therefore using low levels of input technology and resulting in positive energy efficiency ratios
- use of local varieties of crops, wild plants, and animals[52]

These commonalities make it possible to identify how the dynamics of traditional systems might lead to general agricultural improvements. For example, it has been observed in studies since at least 1977 that the productivity of small-scale multiple cropping systems is higher than in monocropping systems.[53] This can result from the interaction between crops, where one plant might release nutrients which benefit another.[54] Additionally, as has been established since at least the Irish potato famine, as a general principle, traditional agroecosystems involving a wide variety of cultivars are less vulnerable to catastrophic crop failure because the diversity of crops offers a variety of defenses against vulnerability.[55]

Food production and distribution systems in developed countries are characterized by industrialized methods of food production and processing, global sources and means of supply, and corporate modes of financing and governance. In developing countries, on the other hand, increasing attention has been given to local knowledge in the construction of alternative food networks (AFNs).[56] Feenstra defines AFNs as "rooted in particular places," with the aim to be "economically viable for farmers and consumers, use ecologically sound production and distribution practises, and enhance social equity and democracy for all members of the community."[57] Examples of AFNs include localized food chains,[58] farmers' markets,[59] and community-supported agriculture.[60]

Nygård and Storstad suggest that smaller-scale local foods are often inherently healthier, safer, and more nutritious than those from larger-scale agricultural production systems.[61] Furthermore, AFNs may confer other ecological benefits, including reduced food miles and carbon emissions,[62] although these positive environmental impacts are still up for debate.[63] Political economists warn in their literature against the over-romanticism of localism,[64] but Tregear cautions that AFNs cannot be properly evaluated until the parameters of the institutions to classify and regulate them are agreed upon.[65]

1.4.2 *Local knowledge and climate change adaptation*

Anthropological studies have demonstrated the usefulness of local knowledge suited to the soil and climate of specific ecosystems in crisis situations.[66] Although the value of local knowledge in informing climate change adaptation has been addressed in a number of studies,[67] the International Panel on Climate Change observed in its 2014 Synthesis Report that "indigenous, local and traditional knowledge systems and practices, including indigenous peoples' holistic view of community and environment, are a major resource for adapting to climate change, but these have not been used consistently in existing adaptation efforts."[68] A 2008 Issues Paper for the International Union for the Conservation of Nature (IUCN) pointed out that indigenous peoples of the tropical forest belt have developed specific coping strategies to deal with extreme variations of weather.[69] The Issues Paper identified the following as adaptation strategies:

(a) crop diversification in order to minimize the risk of harvest failure
(b) changes of living area and a variety of movement patterns to deal with climatic variability
(c) change of hunting and gathering periods in response to changing animal migration and fruiting periods

(d) change of varieties and species cultivated to take into account new disease challenges
(e) changes in food storage methods, such as drying or smoking, in response to climate variability and the corresponding variability of the availability of food
(f) changes in food habits, such as reversion to gathering food in the forests during periods of insufficient agricultural production, or changes in the local food exchange economy
(g) forests as source of famine food in case of emergency[70]

1.4.3 Local knowledge and scientific knowledge

According to the high-modernist ideology pushed by development planners throughout the twentieth century, local knowledge was destined to be superseded by scientific knowledge.[71] It was originally envisaged that pursuant to the "green revolution" hybrid varieties and intensified and mechanized production systems would replace traditional cultivars and cultivation practices. However, over time, the paradigm of economic development theory shifted from "top-down" to "bottom-up" approaches, beginning to highlight the need for community-based natural resource management.[72] The 1992 Convention on Biological Diversity (CBD) urged its convention parties to "respect, preserve and maintain knowledge, innovations and practices of indigenous and local communities embodying traditional lifestyles relevant for the conservation and sustainable use of biological diversity."[73] Similarly, the 2001 International Treaty on Plant Genetic Resources for Food and Agriculture conducted by the FAO emphasized in situ conservation and "farmers' rights" to equitable benefit-sharing and participation in decision-making, in recognition of their past and future contributions to plant conservation and development.[74] New approaches to natural resource management received further boosts from the decentralization policies of developing countries.[75]

The characteristic agricultural system in the mountainous and hilly regions of Latin America, Central Africa, and Southeast Asia is "swidden" agriculture, also known as "slash-and-burn" or "shifting" cultivation.[76] During the first decades following World War II, the knowledge and associated natural resource management practices, including swidden agriculture, used by "traditional" or "indigenous" peoples, were considered to be outmoded, inefficient, and environmentally damaging. However, more recently, it has been acknowledged that traditional and indigenous perspectives might be more environmentally sustainable than previously conceived of in the western tradition. By maintaining a mosaic of cultivated plots, farmers are able to harness the natural processes of soil regeneration. For example, Drahos observed that adopting the fire management practices that the Aboriginal peoples of Australia had employed for thousands of years could also mitigate the ferocity of the annual wildfires which afflict Australia.[77]

1.4.4 Local knowledge in the legal discourse

In legal discourse, local agricultural knowledge primarily attracts the attention of intellectual property (IP), environmental, and international lawyers. IP lawyers focus on the knowledge itself and its protection over a broad range of domains, including cultural practices and rights,[78] human rights,[79] the right to food and food security,[80] the right of access to biological resources,[81] and concerns about "biopiracy," which can arise out of imbalances between strong rights for IP holders and weak public benefits for traditional farmers and holders of local knowledge about biodiversity.[82]

IP analyses tend to focus on the role of the IP system in agricultural innovation and food security, particularly the roles of DNA patenting[83] and plant variety protection[84] in securing investment in agricultural research. However, the most relevant category of IP for the protection and utilization of local knowledge in food and food security are geographical indications (GIs).

1.5 Geographical indications, local knowledge, and food

In accordance with Article 22.2 of the World Trade Organization (WTO) Agreement on Trade-Related Aspects of Intellectual Property Rights (TRIPS), countries are obligated to protect geographical indications. The TRIPS Agreement defines GIs in Article 22.1 as "indications which identify a good as originating in the territory of a Member, or a region or locality in that territory, where a given quality, reputation or other characteristic of the good is essentially attributable to its geographical origin." The local knowledge of people in a particular region, such as their agricultural methods, can provide the decisive linkage between the quality or characteristics of a product and its place of origin.[85]

GIs are particularly advantageous for the producers of agricultural products because they allow them to differentiate their products from general commodity products such as rice, coffee, and tea, thereby enhancing market access.[86] A number of researchers have furthermore identified the capacity of GIs to capture premium prices because of the higher value that some consumers attach to products due to their differentiated origins. For example, Babcock reported that Bresse poultry in France received quadruple the commodity price for poultry meat;[87] a case study by Gerz and Dupont of Comté cheese in France indicated that French farmers receive an average of 14 percent more for milk destined for Comté and that dairy farms in the defined Comté area since 1990 are 32 percent more profitable than similar farms outside the area.[88] Kireeva et al., examining the use of origin marks in the People's Republic of China, reported that the price of "Zhangqiu Scallion" per kilogram was raised from 0.2–0.6 yuan before the use of the origin mark to 1.2–5 yuan in 2009.[89] "Jianlian" lotus seed was registered as a GI in 2006, leading to a rise in price from 26–28 yuan per kilogram to 32–34 yuan per kilogram.

Moreover, GIs can play an important role in signaling the quality of goods to consumers.[90] They are important for communicating credence attributes, particularly as an origin brand will be underpinned by a registration and certification system. Producers can signal quality and the associated reputation that has been developed over time,[91] and can be incentivized by the premium prices attracted by a GI to maintain product quality.[92]

Of course, in order for the perceived benefits of GI labelling such as the promotion of environmental sustainability to be realized there must be consumer awareness that origin labelling represents particular qualities linked to both natural and human factors. This awareness is growing, tied to the current rising consumer demand for traceability in agrifood products.[93] Rural product certification schemes, which have proliferated since the mid-1990s, include the certification of organic agriculture, fair trade certification of products from developing countries, and food produced in compliance with sanitary and traceability protocols, among other systems.[94] According to Giraud and Ambelard, consumers have been shown to place an increasing value on the integrity of the food they purchase, which can include the social and environmental standards involved in the production and processing of agrifood products.[95] This trend has been particularly rising in light of recent high-profile food safety crises. Although it is not unusual for food to be grown, processed, and packaged in different places, consumer trust in such products has eroded, partially as a consequence of these crises. Studies indicate that consumers are willing to pay a premium price to producers who offer transparency about the composition and origin of their products. In situations where uncertainty about quality or safety is elevated, such as in a health crisis, origin labelling can become an important means of inferring product quality. Some examples include meat labels after the BSE crisis in Europe[96] and dairy product labels after the Chinese Melamin crisis.[97]

Concerns about the safety of agrifoods in China has stimulated an interest in mechanisms for assuring traceability in food chains.[98] In this context, GIs "may convey assumed 'locality' (i.e. traceability), and 'natural' (i.e. nutritiousness and safety) characteristics which act as proxies for quality."[99]

In Europe, where GIs have been longest developed, there are some empirically based suggestions that both consumers and producers have expectations about the quality of origin products in the European market.[100] However, studies indicate that although historically European producers have not necessarily specifically addressed positive environmental effects in how they formulate product specifications, more recently the "greening" of product specifications reflecting environmental considerations has been on the rise.[101] Thus, according to Belletti et al., GIs can "provide the opportunity for territorialization of environmentally friendly production rules, taking into account local specificities."[102]

The evolution of specifications of origin products is the result of long-standing farming practices composed of agricultural, cultural, and environmental practices.[103] Traditional crop-management practices have been identified as a rich resource for understanding the interactions between biodiversity and ecosystem function, which is crucial for the identification of principles needed to develop more sustainable agricultural systems.[104]

Furthermore, biodiversity objectives are often incorporated into the codes of practices collectively adopted by producer associations that control origin labels.[105] Biénabe et al. refer to the Rooibos industry in South Africa as an example of an industry which has explicitly considered biodiversity concerns in designing its product specifications, since Rooibos production takes place in a biodiverse and environmentally sensitive area.[106]

With greater knowledge of the interdependence between agricultural products and the local environment, producer associations also have a greater awareness of threats that certain production practices can have on the environment.[107] Consequently, Belletti et al. suggest that the "GI registration process can be expected to have a positive impact upon the key components of ecological embeddedness and, in particular, on the way actors involved in the chain address the ecological elements of food production … ".[108] Kop et al. illustrate this phenomenon with the example of the production of the French cheese Comté, which must follow the specifications agreed upon by its Protected Designation of Origin (PDO) producer association. These specifications limit the intensification of the cheese production, so the cheesemakers use fewer inputs and the environment is better protected, which helps to maintain the open landscape of both pasture and woodland characteristic of the Jura region.[109] The profitability of traditional livestock raising in the Comté geographic area resulting from the price premium attached to the PDO label has limited the loss of pastureland to 7 percent in the GI-approved area, compared with 18 percent in the non-GI area.

Belletti et al., in their empirical study of the European olive oil industry, which is characterized by the extensive use of GIs, identify the industry as a good example of an agriculture system with many associated positive environmental impacts, such as lower rates of soil erosion, improved fire-risk control, increased water efficiency, lower pollution, higher levels of biodiversity, and greater genetic diversity in the olive-tree varieties themselves.[110]

Lamarque and Lambin, in a study of cheese producers in the French Alps, marketing their cheese as "Tomme de Savoie" and "Emmental de Savoie,"[111] found that farmers used GIs to attract price premiums and generally adopted environmentally sustainable cropping practices.[112] However, they conceded that the data from this study might be skewed by the effect of product subsidies under the European Common Agricultural Policy.

1.5.1 Preserving biodiversity through the protection of geographical indications

Rural sustainability achieved through the preservation of biodiversity, landscapes, and traditional knowledge may be promoted by the protection of GIs.[113] For example, Guerra has observed that in the Mexcal region of Mexico, the Agave sugar needed to make Tequila is cultivated and managed from wild or forest Agave species, encouraging the biodiversity of the Agave species.[114] GIs can also serve as a tool for encouraging sustainable agricultural practice by legally limiting the production scale and methods. Penker notes that origin products impose an increased responsibility of producers to their place of production.[115] Lampkin et al. have noted that "organic standards provide a mechanism by which farmers pursuing sustainability goals can be compensated by the market for internalizing external costs."[116]

Bérard and Marchenay describe GIs as a means of "enabling people to translate their long-standing, collective, and patrimonial knowledge into livelihood and income" that may also underpin the maintenance of biodiversity.[117] A number of authors have pointed out that GIs share many of the characteristics of traditional knowledge (TK), as both seek to preserve communal rights and, like traditional knowledge, GIs can be held in perpetuity, for as long as a community maintains the practices which guarantee the distinctive quality of a local product.

In the early days of GI systems a justification advanced for the establishment of a protection system for wines produced in France was the role that their production played in preserving agriculture and rural employment in areas which were unsuitable for cereals and other crops.[118] The maintenance and promotion of rural development have been repeatedly advanced as a justification for GIs.[119]

The creation of local jobs through the protection of GIs is a factor in minimizing rural exodus.[120] An increase in employment has, for example, been observed in the Comté cheese industry. Kop et al. estimate that the production of Comté generates five times more jobs in processing, maturing, marketing, packing, etc. than does its generic equivalent, Emmental, and that migration away from the countryside in the Comté area is only half of that of the origin-protected area.[121] They estimate that, at the national level, although Comté cheeses account for only 10 percent of the total French cheese output, they are responsible for 40 percent of the job offers for students who have been trained in cheese-making in vocational schools. Similar results have been identified for origin-protected cheeses supporting the milk supply industries from cattle in Northern Italy and the sheep of Southern Italy.[122]

1.6 Editors' note

The first question answered in this book focuses on what we know about promoting environmental resilience through agroecology and biodiversity. It seems that local agricultural knowledge (LAK) offers much insight into what we, in fact, collectively can call knowledge that is useful for the protection of the environment and the future of resilient food production. The systematic review of the literature on LAK by Professor Blakeney in this chapter provides an introduction to the multitude of benefits for protecting LAK. Thus, protecting GIs is a tool for preserving LAK.

1.7 EU policy on geographical indications

One of the tools that we have in order to protect local knowledge in food law are, as Blakeney observes, geographic indications (GIs). The European Commission (EC) defines a GI a "a distinctive sign used to identify a product whose quality, reputation or other characteristic is linked to its geographical origin."[123] In fact, the EU has a special policy for the protection of GIs and "is active in multilateral and bilateral negotiations protecting" the EU's GIs. Specifically, in the EU:

> EU quality policy aims at protecting the names of specific products to promote their unique characteristics, linked to their geographical origin as well as traditional know-how. Product names can be granted with a 'geographical indication' (GI) if they have a specific link to the place where they are made. The GI recognition enables consumers to trust and distinguish quality products while also helping producers to market their products better.[124]

Under this general purview, there are three types of designations, as shown in Table 1.1. The European Commission maintains a registry of all the GIs.[125]

Table 1.1 Geographical indications in the EU

	Protected designation of origin (PDO)	Protected geographical indication (PGI)	Geographical indication of spirit drinks and aromatized wines (GI)
Definition	Product names registered as PDO are those that have the strongest links to the place in which they are made.	PGI emphasizes the relationship between the specific geographical region and the name of the product where a particular quality, reputation, or other characteristic is essentially attributable to its geographical origin.	The GI protects the name of a spirit drink or aromatized wine originating in a country, region, or locality where the product's particular quality, reputation, or other characteristic is essentially attributable to its geographical origin.
Products	Food, agricultural products, and wines.	Food, agricultural products, and wines.	Spirit drinks and aromatized wines.
Specifications	Every part of the production, processing, and preparation process must take place in the specific region. For wines, this means that the grapes have to come exclusively from the geographical area where the wine is made.	For most products, at least one of the stages of production, processing, or preparation takes place in the region. In the case of wine, this means that at least 85 percent of the grapes used have to come exclusively from the geographical area where the wine is actually made.	For most products, at least one of the stages of distillation or preparation takes place in the region. However, raw products do not need to come from the region.
Example	Kalamata olive oil PDO is entirely produced in the region of Kalamata in Greece, using olive varieties from that area.	Westfälischer Knochenschinken PGI ham is produced in Westphalia using age-old techniques, but the meat used does not originate exclusively from animals born and reared in that specific region of Germany.	Scotch Whisky GI has been produced for over 500 years in Scotland, including the distillation and maturation, but the raw materials do not come exclusively from Scotland.

(Continued)

Table 1.1 (Continued) Geographical indications in the EU

	Protected designation of origin (PDO)	Protected geographical indication (PGI)	Geographical indication of spirit drinks and aromatized wines (GI)
Label	• mandatory for food and agricultural products • optional for wine	• mandatory for food and agricultural products • optional for wines	• optional for all products

Notes

1. FAO, *The State of Food Insecurity in the World—Meeting the 2015 International Hunger Targets: Taking Stock of Uneven Progress* (Rome: Food and Agriculture Organization, 2015).
2. It should be noted that there is a large number of definitions of food security. Some 200 definitions of food security were noted by S. Maxwell and M. Buchanan-Smith, "Household food security: a conceptual review," in S. Maxwell and T. Frankenburger (eds), *Household Food Security: Concepts, Indicators, Measurements, A Technical Review* (New York and Rome: UNICEF and UNCTAD, 1992). I. Scoones, "Agricultural Biotechnology and Food Security: Exploring the Debate," IDS working paper 145 (January 2002) traces the definition from the 1974 World Food Conference connotation of access to the availability of food (referring to UN, Report of the World Food Congress, New York, November 5–16, 1974). Indicators of food security can be defined at different levels: for the world as a whole, for individual countries, or for households. At the national level, adequate food availability means that on average sufficient food supplies are available from domestic production and/or imports to meet the consumption needs of all in the country (see FAO, "Some issues relating to food security in the context of the WTO negotiations on agriculture," FAO Geneva Round Table on Food Security in the Context of the WTO Negotiations on Agriculture, July 20, 2001).
3. See Stephen A. Wood, Amir S. Jina, Meha Jain, Patti Kristjanson, and Ruth S. DeFries, "Smallholder farmer cropping decisions related to climate variability across multiple regions" (2014) 25 *Global Environmental Change* 163.
4. FAO, Committee on World Food Security, Impact of Climate Change on Food Security and Implications for Sustainable Food Production, May 12, 2003, FAO, Rome, Doc. CFS:2003/INF.
5. FAO, Global Climate Change and Agricultural Production: Direct and Indirect Effects of Changing Hydrological, Pedological, and Plant Physiolocial Processes, CFS:2003/INF/11.
6. P. Pinstrup-Andersen, R. Pandya-Lorch, and M.W. Rosegrant, *World Food Prospects: Critical Issues for the Early Twenty-First Century* (Washington, DC: International Food Policy Research Institute, 1999), chapter 1.
7. I. Serageldin and G.J. Pursley, *Promethean Science. Agricultural Biotechnology, the Environment and the Poor* (Washington, DC: CGIAR, 2000), 3.
8. See G. Conway and J. Pretty, *Unwelcome Harvest. Agriculture and Pollution* (London: Earthscan, 1991).
9. See G. Conway, *The Doubly Green Revolution—Food for All in the Twenty-First Century* (Harmondsworth: Penguin, 1997).
10. See e.g. C. Thorburn, "The rise and demise of integrated pest management in rice in Indonesia" (2015) 6 *Insects* 381.
11. See J. Pretty, *Regenerating Agriculture. Policies and Practices for Sustainability and Self-Reliance* (London: Earthscan, 1995).
12. See W.T. Winarto, K. Stigter, B. Dwisatrio, M. Nurhaga, and A. Bowolaksono, "Agrometeorological learning increasing farmers' knowledge in coping with climate change and unusual risks" (2013) 2 *Southeast Asian Studies* 323.
13. M. Blakeney, "The negotiations in WIPO for international conventions on traditional knowledge and traditional cultural expressions," in Jessica C. Lai

and Antoinette Maget Dominicé (eds), *Intellectual Property and Access to Im/material Goods* (Cheltenham: Edward Elgar, 2016), 227–256.

14. WIPO Secretariat, "Glossary of Key Terms Related to Intellectual Property and Genetic Resources, Traditional Knowledge and Traditional Cultural Expressions," 21st Session, WIPO Doc. WIPO/GRTKF/IC/28/INF/7, 2014.
15. Ibid., at 40.
16. Citing WIPO, *Intellectual Property Needs and Expectations of Traditional Knowledge Holders WIPO, WIPO Report on Fact-finding Missions on Intellectual Property and Traditional Knowledge (1998–1999)*, (Geneva: WIPO, 2001), 25.
17. Erica-Irene Daes, "Discrimination against indigenous peoples," E/CN.4/Sub.2/1993/28, July 28, 1993, para. 4.
18. *United Nations Declaration on the Rights of Indigenous Peoples*, GA Res 61/295, UN GAOR, 61st sess., 107th plen. mtg., Supp No 49, UN Doc A/RES/61/295, September 13, 2007, art. 31.
19. Sharon B. Le Gall, *Intellectual Property, Traditional Knowledge and Cultural Property Protection. Cultural Signifiers in the Caribbean and the Americas* (London: Routledge, 2014).
20. M.R. Dove, "The life-cycle of indigenous knowledge, and the case of natural rubber production," in R. Ellen, P. Parkes, and A. Bicker (eds), *Indigenous Environmental Knowledge and its Transformations* (Amsteldijk: Harwood Press, 2000), 213 at 214–215.
21. B. Kingsbury, "The applicability of the international legal concept of 'indigenous peoples' in Asia," in J.R. Bauer and D.A. Bell (eds), *The East Asian Challenge for Human Rights* (Cambridge: Cambridge University Press, 1999), 336–377; Li T. Murray, "Locating indigenous environmental knowledge in Indonesia," in Ellen, Parks, and Bicker (eds.) *Indigenous Environmental Knowledge and Its Transformations: Critical Anthropological Perspectives*, Routledge, London, Routledge, 2000, 121–149; G. Benjamin, "On being tribal in the Malay world," in G. Benjamin and C. Chou (eds), *Tribal Communities in the Malay World: Historical, Cultural and Social Perspectives* (Leiden: Institute of Southeast Asian Studies and International Institute for Asian Studies, Singapore, 2002), 7–76; G.A. Persoon, "'Being indigenous' in Indonesia and the Philippines," in C. Antons (ed.), *Traditional Knowledge, Traditional Cultural Expressions and Intellectual Property Law in the Asia-Pacific Region* (Alphen aan den Rijn: Kluwer Law International, 2009), 195–216; C. Antons, "Traditional knowledge, biological resources and intellectual property rights in Asia: the example of the Philippines" (2007) 34 *Forum of International Development Studies* 1.
22. See C. Antons, "Traditional cultural expressions and their significance for development in a digital environment: examples from Australia and Southeast Asia," in C.B. Graber and M. Burri-Nenova (eds), *Intellectual Property and Traditional Cultural Expressions in a Digital Environment* (Cheltenham: Edward Elgar, 2008), 287–301 at 295; T. Forsyth and A Walker, *Forest Guardians, Forest Destroyers: The Politics of Environmental Knowledge in Northern Thailand* (Chiang Mai: Silkworm Books, 2008), 213–214.
23. P. Sillitoe, "Ethnobiology and applied anthropology: *rapprochement* of the academic with the practical," in R. Ellen (ed.), *Ethnobiology and the Science of Humankind* (Oxford: Blackwell Publishing, 2006), 147–175.
24. D. Frossard, "Peasant science: a new paradigm for sustainable development?" (1998) 17 *Research in Philosophy and Technology* 111; Y. T. Winarto,

Seeds of Knowledge: The Beginnings of Integrated Pest Management in Java (New Haven, CT: Yale University Southeast Asian Studies, 2004).

25. M.R. Dove, "Use of global legal mechanisms to conserve local biogenetic resources: problems and prospects," in M.R. Dove, P.E. Sajise, and A.A. Doolittle (eds), *Conserving Nature in Culture: Case Studies from Southeast Asia* (New Haven, CT: Yale University Southeast Asian Studies, 2005), 279–305; Sillitoe, "Ethnobiology and applied anthropology."
26. R.A. Cramb, *Land and Longhouse: Agrarian Transformation in the Uplands of Sarawak* (Copenhagen: NIAS Press, 2007).
27. M.R. Dove, "The life-cycle of indigenous knowledge, and the case of natural rubber production," in Ellen et al., *Indigenous Environmental Knowledge,* 213–251.
28. WIPO Secretariat, "Traditional knowledge: operational terms and definitions," 3rd Session, WIPO Doc. WIPO/GRTKF/IC/3/9, 2002, para. 4.
29. Ibid., para. 5.
30. P. Drahos, *Intellectual Property, Indigenous People and their Knowledge* (Cambridge: Cambridge University Press, 2014), 203.
31. See A. Peukert, "Intellectual property: the global spread of a legal concept," in P. Drahos, G. Ghidini, and H. Ullrich (eds), *Kritika: Essays on Intellectual Property,* vol. 1 (Cheltenham: Edward Elgar, 2015), 84–113.
32. M. Forsyth, "Making room for magic in intellectual property policy," ibid. et al., 114–133.
33. S. Bragdon, "Small scale farmers: the missing element in the WIPO–IGC draft articles on genetic resources," *QUNO Briefing Paper* 1 (2013).
34. See B.M. Tobin, "Bridging the Nagoya compliance gap: the fundamental role of customary law in protection of indigenous peoples" (2013) 9(2) *Law, Environment and Development Journal* 142.
35. See M. Forsyth, "How can the theory of legal pluralism assist the traditional knowledge debate?" (2013) *Intersections: Gender and Sexuality in Asia and the Pacific* 33.
36. See M. Blakeney, "The negotiations in WIPO for international conventions on traditional knowledge and traditional cultural expressions," in Lai and Dominicé, *Intellectual Property,* 227–256.
37. See M. B. Jensen, B. Johnson, E. Lorenz, and B.A. Lundvall, "Forms of knowledge and modes of innovation" (2007) 36 *Research Policy* 680.
38. See J. Charmes, "The informal economy: definitions, size, contribution and main characteristics," in Erika Kraemer-Mbula and Sacha Wunsch-Vincent (eds), *The Informal Economy in Developing Nations: Hidden Engine of Innovation* (Cambridge: Cambridge University Press, 2016), 13–42.
39. E.g. K. Hart, "Informal income opportunities and urban employment in Ghana" (1973) 11(1) *Journal of Modern African Studies* 61.
40. ILO, *Employment, Incomes and Equality: A Strategy for Increasing Productive Employment in Kenya* (Geneva: International Labour Office, 1973), 6.
41. B. Hall, C. Helmers, M. Rogers, and V. Sena, "The choice between formal and informal intellectual property: a review" (2014) 52(2) *Journal of Economic Literature* 1.
42. J. de Beer and S. Wunsch-Vincent, "Appropriation and intellectual property in the informal economy," in Kraemer-Mbula and Wunsch-Vincent, *The Informal Economy,* 232–276 at 239.

43. Ibid., at 237.
44. See L. Klerkz and P. Gildemacher, "The role of innovation brokers in agricultural innovation systems," in World Bank, *Agricultural Innovation Systems: An Investment Sourcebook* (Washington, DC, World Bank, 2012), 211–230.
45. See Y.T. Winarto, Kees Stigter, Bimo Dwisatrio, Merryna Nurhaga, and Anom Bowolaksono, "Agrometeorological learning increasing farmers' knowledge in coping with climate change and unusual risks" (2013) 2(2) *Southeast Asian Studies* 323.
46. See W.M. Denevan, "Prehistoric agricultural methods as models for sustainability" (1995) 11 *Advances in Plant Pathology* 21.
47. Roy Ellen and Hermien L. Soselisa, "A comparative study of the socio-ecological concomitants of cassava (*Manihot esculenta* Crantz) diversity, local knowledge and management in Eastern Indonesia" (2012) 10 *Ethnobotany Research and Applications* 15.
48. M.A. Altieri, *Biodiversity and Pest Management in Agroecosystems* (New York: Haworth Press, 1994); J.P Rosenthal and R. Dirzo, "Effects of life history, domestication and agronomic selection on plant defense against insects: evidence from maizes and wild relatives" (1997) 11 *Evolutionary Ecology* 337.
49. The material in this sentence is taken from a paper published by Michael Blakeney in June 2017: M. Blakeney, "Food safety and free trade: geographical indications and environmental protection" (2017) 12 *Frontiers of Law in China* 2.
50. B.R. Dewalt, "Using indigenous knowledge to improve agriculture and natural resource management" (1994) 5 *Human Organization* 23.
51. M.A. Altieri, "Linking ecologists and traditional farmers in the search for sustainable agriculture" (2004) 2 *Frontiers in Ecology and the Environment* 35 at 36, citing S.R. Gliessman, *Agroecology: Ecological Processes in Sustainable Agriculture* (Chelsea, MI: Ann Arbor Press, 1998).
52. Ibid.
53. J.H. Chang, "Tropical agriculture: crop diversity and crop yields" (1977) 53 *Economic Geography* 24.
54. J. Vandermeer, *The Ecology of Intercropping* (Cambridge: Cambridge University Press, 1989).
55. Lori Thrupp, *Cultivating Diversity: Agrobiodiversity for Food Security* (Washington, DC: World Resources Institute, 1998).
56. See David Goodman and Michael K. Goodman, *Alternative Food Networks: Knowledge, Practice, and Politics* (Hoboken, NJ: Taylor & Francis, 2012).
57. G. Feenstra, "Local food systems and sustainable communities" (1997) 12 *American Journal of Alternative Agriculture* 28–36.
58. T. Marsden, J. Banks, and G. Bristow, "Food supply chain approaches: exploring their role in rural development" (2000) 40 *Sociologia Ruralis* 424; C. Hinrichs, "Embeddedness and local food systems: notes on two types of direct agricultural market" (2000) 16 *Journal of Rural Studies* 295; H. Renting, T. Marsden, and J. Banks, "Understanding alternative food networks: exploring the role of short food supply chains in rural development" (2003) 35 *Environment and Planning* 393; B. Ilbery and D. Maye, "Alternative (shorter) food supply chains and specialist livestock products in the Scottish–English borders" (2005) 37 *Environment and Planning* 823.
59. L. Holloway and M. Kneafsey, "Reading the space of the farmers' market: a preliminary investigation from the UK" (2000) 40 *Sociologia Ruralis* 285; J.

Kirwan, "The interpersonal world of direct marketing: examining conventions of quality at UK farmers' markets" (2006) 22 *Journal of Rural Studies* 301; C. Brown and S. Miller, "The impacts of local markets: a review of research on farmers' markets and community supported agriculture" (2008) 90 *American Journal of Agricultural Economics* 1296.

60. P. Allen, M. FitzSimmons, M. Goodman, and K. Warner, "Shifting places in the agrifood landscape: the tectonics of alternative agrifood initiatives in California" (2003) 19 *Journal of Rural Studies* 61.
61. B. Nygård and O. Storstad, "De-globalization of food markets? Consumer perceptions of safe food: the case of Norway" (1998) 38 *Sociologia Ruralis* 35; L. Cembalo, A. Lombardi, S. Pascucci, D. Dentoni, et al., "Rationally local: consumer participation in alternative food chains" (2015) 31 *Agribusiness* 330; Stefano Pascucci, Domenico Dentoni, Alessia Lombardi, and Luigi Cembalo, "Sharing values or sharing costs? Understanding consumer participation in alternative food networks" (2016) 78 *NJAS Wageningen Journal of Life Sciences*, 470; Filippo Barbera and Joselle Dagnes, "Building alternatives from the bottom-up: the case of alternative food networks" (2016) 8 *Agriculture and Agricultural Science Procedia* 324.
62. M. Kneafsey, R. Cox, L. Holloway, E. Dowler, et al., *Reconnecting Consumers, Producers and Food: Exploring Alternatives* (Oxford: Berg, 2008).
63. See G. Edwards-Jones, Llorenç Milà i Canals, Natalia Hounsome, Monica Truninger et al., "Testing the assertion that 'local food is best': the challenges of an evidence-based approach" (2008) 19 *Trends in Food Science and Technology* 265; D. Oglethorpe, "Food miles: the economic, environmental and social significance of the focus on local food" (2009) 4 *CAB Reviews: Perspectives in Agriculture, Veterinary Science, Nutrition and Natural Resources* 72.
64. M. DuPuis and D. Goodman, "Should we go 'home' to eat? Toward a reflexive politics of localism" (2005) 21 *Journal of Rural Studies* 359; M. DuPuis, D. Goodman, and J. Harrison, "Just values or just value? Remaking the local in agro-food studies," in Terry Marsden and Jonathan Murdoch (eds), *Between the Local and the Global Research in Rural Sociology and Development*, vol. 12 (London: Emerald Group Publishing, 2006), 241–268; A. Lombardi, G. Migliore, F. Verneau, G. Schifani, and L. Cembalo, "Are 'good guys' more likely to participate in local agriculture?" (2015) 45 *Food Quality and Preference* 158.
65. See Angela Tregear, "Progressing knowledge in alternative and local food networks: critical reflections and a research agenda" (2011) 27 *Journal of Rural Studies* 419.
66. F. Berkes, J. Colding, and C. Folke, "Rediscovery of traditional ecological knowledge as adaptive management" (2000) 10 *Ecological Applications* 1251; R. Ellen, Introduction, in R. Ellen (ed.), *Modern Crisis and Traditional Strategies*, Local Ecological Knowledge in Island Southeast Asia (New York, Berghahn Books, 2007), 1; J. Salick and N. Ross, "Traditional peoples and climate change" (2009) 19 *Global Environmental Change* 137; D. Green and G. Raygorodetsky, "Indigenous knowledge of a changing climate" (2010) 100 *Climatic Change* 239; Karen Elizabeth McNamara and Ross Westoby, "Local knowledge and climate change adaptation on Erub Island, Torres Strait" (2011) 16 *Local Environment* 887.
67. D. Riedlinger and F. Berkes, "Contributions of traditional knowledge to understanding climate change in the Canadian Arctic" (2001) 37 *Polar Record*

315; M.S Reed, A.J. Dougill, and M.J. Taylor, "Integrating local and scientific knowledge for adaptation to land degradation: Kalahari range-land management options" (2007) 18 *Land Degradation and Development* 249; L.C. Stringer, J.C. Dyer, M.S Reed, A.J. Dougill, et al., "Adaptations to climate change, drought and desertification: local insights to enhance policy in southern Africa" (2009) 12 *Environmental Science and Policy* 748; Green and Raygorodetsky, "Indigenous knowledge"; Andrew J. Newsham and David S.G. Thomas, "Knowing, farming and climate change adaptation in North-Central Namibia" (2011) 21 *Global Environmental Change* 761.

68. IPCC (International Panel on Climate Change), "Climate Change 2014," Synthesis Report, WHO and Gland UNEP, Geneva, 2014, at para 4.4.2.
69. Mirjam Macchi, Gonzalo Oviedo, Sarah Gotheil, Katharine Cross et al., "Indigenous and traditional peoples and climate change," Issues Paper, International Union for the Conservation of Nature, Gland, 2008.
70. Ibid., at 40–41.
71. R. Cramb, *Land and Longhouse*, 28; R. Ellen, Introduction, 6; J.C. Scott, *Seeing Like a State: How Certain Schemes to Improve the Human Condition Have Failed* (New Haven, CT: Yale University Press, 1998).
72. T.M. Li, *The Will to Improve: Governmentality, Development and the Practice of Politics* (Durham, NC: Duke University Press, 2007); J.P. Brosius, A. Lowenhaupt Tsing, and C. Zerner, *Communities and Conservation: Histories and Politics of Community-Based Natural Resource Management* (Lanham, MD: Altamira Press, 2005).
73. Article 8j CBD.
74. S.B. Brush, "Farmers' rights and protection of traditional agricultural knowledge" (2007) 35 *World Development* 1499.
75. C. Wittayapak and P. Vandergeest, *The Politics of Decentralization: Natural Resource Management in Asia* (Chiang Mai: Mekong Press, 2010).
76. J.G. Goldjammer, "Rural land-use and wildland fires in the tropics" (1988) 6 *Agroforest Systems* 235; V.N. Meine, M. Elok, S. Niken, and A. Fahmuddin, *Swiddens in Transition: Shifted Perceptions on Shifting Cultivators in Indonesia* (Bogor, Indonesia: World Agroforestry Centre, 2008); O. Merz, Christine Padoch, Jefferson Fox, R. A. Cramb et al., "Swidden change in Southeast Asia: understanding causes and consequences" (2009) 37 *Human Ecology* 339; J. Fox, Yayoi Fujita, Dimbab Ngidang, Nancy Peluso et al., "Policies, political-economy and swiddens in Southeast Asia" (2009) 37 *Human Ecology* 305; A. D. Ziegler et al., "Recognizing contemporary roles of swidden agriculture in transforming landscape of Southeast Asia" (2010) 25 *Conservation Biology* 846; N. Van Vliet, "Trends, drivers and impacts of changes in swidden agriculture in tropical forest-agriculture frontiers: a global assessment" (2012) 22 *Global Environmental Changes* 418; Peng Li, Zhiming Feng, Luguang Jiang, Chenhua Liao, and Jinghua Zhang, "A review of swidden agriculture in Southeast Asia" (2014) 6 *Remote Sensing* 1654.
77. P. Drahos, *Intellectual Property, Indigenous People and Their Knowledge* (Cambridge: Cambridge University Press, 2014).
78. S. von Lewinski, *Indigenous Heritage and Intellectual Property: Genetic Resources, Traditional Knowledge and Folklore*, 2nd ed. (Alphen aan den Rijn: Wolters Kluwer, 2008); J. Gibson, *Community Resources: Intellectual Property, International Trade, and the Protection of Traditional Knowledge* (London: Ashgate

Publishing, 2005); L. Lixinski, *Intangible Cultural Heritage in International Law* (Oxford: Oxford University Press, 2013.

79. L.R. Helfer and G.W. Austin, *Human Rights and Intellectual Property: Mapping the Global Interface* (Cambridge: Cambridge University Press, 2011).
80. M. Blakeney, *Intellectual Property Rights and Food Security*; B. Sherman, "Reconceptualizing intellectual property to promote food security," in C Lawson and J. Sanderson (eds), *The Intellectual Property and Food Project: From Rewarding Innovation and Creation to Feeding the World* (Farnham: Ashgate 2013), 23–38.
81. Christoph Antons, "The role of traditional knowledge and access to genetic resources in biodiversity conservation in Southeast Asia" (2010) 19 *Biodiversity and Conservation* 1189.
82. See M. Blakeney, "Bioprospecting and biopiracy," in B. Ong (ed.), *Intellectual Property and Biological Resources* (Singapore: Marshall Cavendish, 2004), 393–424; D.F. Robinson, *Confronting Biopiracy: Challenges, Cases and International Debates* (London: Earthscan, 2010).
83. M. Blakeney, "Stimulating agricultural innovation," in K.E. Maskus and J.H. Reichman (eds.), *International Public Goods and Transfer of Technology under a Globalized Intellectual Property Regime* (Cambridge: Cambridge University Press, 2005), 367–390; M. Blakeney, "Patents and plant breeding: implications for food security" (2011) 3 *Amsterdam Law Forum* 73; M. Blakeney, "DNA patenting," in Harikesh B. Singh, Alok Jha, and Chetan Keswani (eds), *Intellectual Property Issues in Biotechnology* (Wallingford: CAB International, Inc., 2016), 128–137.
84. Deepthi Elizabeth Kolady and William Lesser, "Does plant variety protection contribute to crop productivity? Lessons for developing countries from US wheat breeding" (2009) 12 *Journal of World Intellectual Property* 137; M. Blakeney, "Patenting of plant varieties and plant breeding methods" (2012) 63 *Journal of Experimental Botany* 1069; J. Sanderson, "Can intellectual property help feed the world? Intellectual property, the PLUMPYFIELD® network and a sociological imagination," in C. Lawson and J. Sanderson (eds), *The Intellectual Property and Food Project: From Rewarding Innovation and Creation to Feeding the World* (Farnham: Ashgate, 2013), 145–173.
85. The material in this section is taken from Blakeney, "Food safety and free trade."
86. See P. Evans, "Geographic indications, trade and the functioning of markets," in M. Pugatch (ed.), *The Intellectual Property Debate: Perspectives from Law, Economics and Political Economy* (London: Edward Elgar, 2006), 345–360; C. Bramley and E. Bienabe, "Developments and considerations around geographical indications in the developing world" (2012) 2 *Queen Mary Journal of Intellectual Property* 14; F. Galtier, G. Belletti, and A. Marescotti, "Are geographical indications a way to 'decommodify' the coffee market?" Paper presented at by 12th Congress of the European Association of Agricultural Economists, Ghent, Belgium, August 26–29, 2008.
87. B.A. Babcock, "Geographical indications, property rights, and value-added agriculture" (2003) *Iowa Ag Review* 9(4), available at www.card.iastate.edu/iowa_ag_review/fall_03/article1.aspx.
88. A. Gerz and F. Dupont, "Comté cheese in France: impact of a geographical indication on rural development," in P. van de Kop, D. Sautier, and A.

Gerz (eds), *Origin-Based Products: Lessons for Pro-Poor Market Development* (Amsterdam: KIT Publishers, 2006), 75–87. See also F. Arfini, "The value of typical products: the case of Prosciutto di Parma and Parmigiano Reggiano cheese," in B. Sylvander, D. Barjolle, and F. Arfini (eds), "The socio-economics of origin labelled products in agri-food supply chains: spatial, institutional and co-ordination aspects" (2000) 17 *Actes et Communications* 77.

89. I. Kireeva, W. Xiaobing, and Z. Yumin, *Comprehensive Feasibility Study for Possible Negotiations on a Geographical Indications Agreement between China and the EU*, EU-China IP2, Brussels, 2009.
90. See J. E. Hobbs, "Information, incentives and institutions in the agri-food sector" (2003) 51 *Canadian Journal of Agricultural Economics* 413; J. E. Hobbs and W.A. Kerr, "Consumer information, labelling and international trade in agri-food products" (2006) 31 *Food Policy* 78; T. Becker, "European food quality policy: the importance of geographical indications, organic certification and food quality insurance schemes in European countries" (2008) 10 *The Estey Centre Journal of International Law and Trade Policy* 111.
91. See J.A Winfree and J. McCluskey, "Collective reputation and quality" (2005) 87 *Journal of Agricultural Economics* 206.
92. G. Moschini, L Menapace, and D Pick, "Geographical indications and the provision of quality in agricultural markets" (2008) 90 *American Journal of Agricultural Economics* 794.
93. J. Murdoch, T. Marsden, and J. Banks, "Quality, nature, and embeddedness: some theoretical considerations in the context of the food sector" (2000) 76(2) *Economic Geography* 107; J.D. Van Der Ploeg, H. Renting, and M. Minderhoud-Jones, "The socio-economic impact of rural development: realities and potentials" 40 *Sociologia Ruralis* 391.
94. G. Giraud and C. Ambelard, "What does traceability mean for beef meat consumer?" (2003) 23 *Food Science* 40; T. Mutersbaugh, D. Klooster, M.-C. Renard, and P. Taylor, "Editorial. Certifying rural spaces: Quality certified products and rural governance" (2005) 21 *Journal of Rural Studies* 381.
95. Giraud and Ambelard, "What does traceability mean"; H. Renting, T. K. Marsden, and J. Banks, "Understanding alternative food networks: exploring the role of short food supply chains in rural development" (2003) 35 *Environment and Planning A* 393; J.E. Hobbs, D. Bailey, D.L. Dickinson, and M. Haghiri, "Traceability in the Canadian red meat sector: do consumers care?" (2005) 53 *Canadian Journal of Agricultural Economics* 47.
96. W. Verbeke and J. Viaene, "Consumer attitude to beef quality labeling and associations with beef quality labels" (1999) 10 *Journal of International Food and Agribusiness Marketing* 45; M.L. Loureiro and W. J. Umberger, "A choice experiment model for beef: what US consumer responses tell us about relative preferences for food safety, country-of-origin labeling and traceability" (2007) 32 *Food Policy* 496; T. Becker, "European food quality policy: the importance of geographical indications, organic certification and food quality insurance schemes in European countries" (2008) 10 *Estey Centre Journal of International Law and Trade Policy* 111; M. Lees, *Food Authenticity and Traceability* (Cambridge: Woodhead Publishing, 2003).
97. L. Xu and L. Wu. "Food safety and consumer willingness to pay for certified traceable food in China" (2010) 90 *Journal of the Science of Food and Agriculture* 1368.

98. See Xing Zhao, Donald Finlay and Moya Kneafsey, "The effectiveness of contemporary geographical indications (GIs) schemes in enhancing the quality of Chinese agrifoods: experiences from the field" (2014) 36 *Journal of Rural Studies* 77.
99. Ibid., at 78.
100. See R. Teuber, "Consumers' and producers' expectations towards geographical indications: empirical evidence for a German case study" (2011) 113 *British Food Journal* 900; A. Stasi, G. Nardone, R. Viscecchia, and A. Seccia, "Italian wine demand and differentiation effect of geographical indications" (2011) 23 *International Journal of Wine Business Research* 49.
101. See D. Giovannucci, T. Josling, W. Kerr, B. O'Connor, and M.T. Yeung, *Guide to Geographical Indications: Linking Products and Their Origin* (New York: International Trade Center, 2009).
102. G. Belletti, A. Marescotti, Javier Sanz-Cañada, and Hristos Vakoufaris, "Linking protection of geographical indications to the environment: Evidence from the European Union olive-oil sector" (2015) 48 *Land Use Policy* 94.
103. See W.M. Denevan, "Prehistoric agricultural methods as models for sustainability" (1995) 11 *Advances in Plant Pathology* 21.
104. B.R. Dewalt, "Using indigenous knowledge to improve agriculture and natural resource management" (1994) 5 *Human Organization* 23.
105. J. Larson, *The Relevance of Geographical Indications and Designations of Origin for the Sustainable Use of Genetic Resources* (Rome: Global Facilitation Unit for Underutilised Species, 2007).
106. E. Biénabe, M. Leclercq, and P. Moity Maizi, "Le rooibos d'Afrique du Sud: comment la biodiversité s'invite dans la construction d'une indication géographique" (2009) 50 *Autrepart* 50.
107. See M. Riccheri, G. A. Benjamin, S. Schlegel, and A. Leipprand, "Assessing the applicability of geographical indications as a means to improve environmental quality in affected ecosystems and the competitiveness of agricultural products," IPDEV Research Report, European Commission, Brussels, 2007.
108. See Belletti et al., "Linking protection of geographical indications" n.63 supra at 95; T. Marsden, J. Banks and G. Bristowet al., "Food supply chain approaches: exploring their role in rural development" (2000) 40 *Sociologica. Ruralis* 424.
109. P. van de Kop, D. Sautier, and A. Gerz, "Origin-based products: lessons for pro-poor market development," Bulletin 372, Royal Tropical Institute (KIT), Amsterdam, French Agricultural Research Centre for International Development (CIRAD), Montpellier, 2006.
110. See Belletti, "Linking protection of geographical indications."
111. Savoie is the name of the Auverne–Rhône-Alpes region of the French Alps.
112. Pénélope Lamarque and Eric F. Lambin, "The effectiveness of marked-based instruments to foster the conservation of extensive land use: the case of geographical indications in the French Alps" (2015) 42 *Land Use Policy* 706.
113. E. Barham, "Towards a theory of value-based labeling" (2002) 19 *Agriculture and Human Values* 349.
114. J. L. Guerra, "Geographical indications and biodiversity: bridges joining distant territories" (2004) *BRIDGES BioRes* 2. See also S. Bowen and A. V.

Zapata, "Geographical indications, terroir, and socio-economic and ecological sustainability: the case of tequila" (2009) 25 *Journal of Rural Studies* 108.

115. M. Penker, "Mapping and measuring the ecological embeddedness of food supply chains" (2006) 37 *Geoforum* 368.
116. N. Lampkin, C. Foster, and S. M. Padel, *The Policy and Regulatory Environment for Organic Farming in Europe: Country Reports*, Organic Farming in Europe: Economics and Policy, vol. 2 (Stuttgart : University of Hohenheim, 1999).
117. L. Bérard and P. Marchenay, "Geographical indications: a contribution to maintaining biodiversity?" Biosphere Reserves, Technical Notes 3-2008, Paris, UNESCO, 2009.
118. A. Stanziani, "Wine reputation and quality controls: the origin of the AOCs in 19th century France" (2004) 18 *European Journal of Law and Economics* 149.
119. C. Ray, "Culture, intellectual property and territorial rural development" (1998) 38 *Sociologia Ruralis* 3; J. Banks and T.K. Marsden, "Integrating agri-environment policy, farming systems and rural development: tir cymen in Wales" (2000) 40 *Sociologia Ruralis* 466; Marsden, et al., "Food supply chain approaches"; B. Ilbery, M. Kneafsey, A. Söderlund, and E. Dimara, "Quality, imagery and marketing: producer perspectives on quality products and services in the lagging rural regions of the European Union" (2001) 83 *Geografiska Annaler* 27; A. Pacciani, G. Beletti, A. Marescotti, and S. Scaramuzzi, "The role of typical products in fostering rural development and the effects of regulation (EEC) 2081/92," 73rd Seminar of the European Association of Agricultural Economists Ancona, June28–30, 2001; G. Belletti and A. Marescotti, "Link between origin-labeled products and rural development," WP Report 3, Development of Origin-Labeled Products: Humanity, Innovations, and Sustainability (DOLPHINS) project, Le Mans, 2002; A. Treagar, "From Stilton to Vimto: using food history to rethink typical products in rural development" (2003) 43 *Sociologia Ruralis* 92; B. Babcock and R. Clemens, "Geographical indications and property rights: protecting value-added agricultural products," MATRIC Briefing paper, 04-MBP 7, 2004; A. Tregear, F. Arfini, G. Belletti, and A. Marescotti, "Regional foods and rural development: the role of product qualification" (2007) 23 *Journal of Rural Studies* 12; G. Belletti and A. Marescotti, "Origin products, geographical indications and rural development," in E. Barham and B. Sylvander (eds), *Labels of Origin for Food: Local Development, Global Recognition* (Wallingford: CAB International, 2011), 75–91; M. Blakeney and G. Mengistie, "Intellectual property and economic development in Sub-Saharan Africa" (2011) 14 *Journal of World Intellectual Property* 238.
120. See O'Connor & Co., "Geographical indications and the challenges for ACP countries," Agritrade, CTA, 2005.
121. Van de Kop, et al., "Origin-based products." See also V. Requillart, "On the economics of geographical indications in the EU," Paper presented at "Workshop on Geographical Indications, Country of Origin and Collective Brands: Firm Strategies and Public Policies," Toulouse, June 14–15, 2007.
122. G. Belletti, G. Brunori, A. Marescotti, and A. Rossi, "Multifunctionality and rural development: a multilevel approach," in G. Van Huylenbroek and G. Durand (eds), *Multifunctional Agriculture: A New Paradigm for European Agriculture and Rural Development* (Aldershot: Ashgate, 2003), 55–80.

123. European Commission, "Geographical indication," http://ec.europa.eu/trade/policy/accessing-markets/intellectual-property/geographical-indications/.
124. European Commission, "Aims of EU quality schemes," https://ec.europa.eu/info/food-farming-fisheries/food-safety-and-quality/certification/quality-labels/quality-schemes-explained_en.
125. The European Commission distinguished between three GI marks: "GIs, PDOs and PGIs protect the name of a product, which is from a specific region and follow a particular traditional production process. However, there are differences between the 3, linked primarily to how much of the raw materials come from the area or how much of the production process has to take place in the specific region."All the information in this table is copied or adapted from https://ec.europa.eu/info/food-farming-fisheries/food-safety-and-quality/certification/quality-labels/quality-schemes-explained_en.

chapter two

Agrobiodiversity, agroecology, and private law

Alberto Giulio Cianci

Contents

2.1 *Private law and agriculture law*

Agrobiodiversity and agroecology represent two important new trends in the field of agriculture law. Agriculture law is an important area in the subdivisions of private law as it includes all regulatory laws related to agricultural activities and to the buying and selling of agricultural products.

The relationship between private law and agriculture law has changed in recent years, especially regarding new developments in agriculture law. Some of the most important changes in this regard are the new functions of agriculture law to ensure the protection of consumers and to give consumers more information. In many legal systems, new regulatory approaches have given to agriculture law added functions in order to ensure consumers have an enhanced level of protection and are given more information concerning their purchases.

As a key part of private law is contract law, and agricultural products are, in this context, the object of a contract of sale, new developments in agriculture law also affect the regulation of consumer contracts. For example, if a new law prescribes mandatory labeling for some agricultural products which were previously exempted from this requirement, it means that this new rule, defined in a specific way for agricultural products, will also have some effects on, and introduce some changes in, contract law.[1] The sale of products will be affected by an additional mandatory requirement, relevant in terms of the specific quality of the goods. This is based on declarations by the producer, which must be truthful, and will have potential effects on consumer choice, mainly by influencing the consumer's decision to complete a sale contract and buy the product.[2]

This new regulatory approach means that the relationship between private law and agriculture law will not follow the traditional pattern, according to which agriculture law is considered simply as a specific area of private law, regulating the activities of businessmen and women in the Italian law system. To take an Italian example: Article 2135 in the Italian Civil Code provides the definition of the agricultural businessman or woman[3] and is considered essential in setting the boundaries of agriculture law as a specific sector of the wider area of private law. The new trend will give to agriculture law other functions, which will in turn lead to integration and interaction with different areas of private law, such as contract law and consumer law.

Because of these developments, the spheres of agriculture law are nowadays greater, and therefore new agriculture laws affect contract law, consumer law, administrative law (with reference to authorizations required for selling products), and many other areas of private law.

2.2 *Agrobiodiversity and agroecology as renewed trends of agriculture law*

Agrobiodiversity and agroecology represent new trends in agriculture law. From a private law perspective, the main question about the importance of considering agrobiodiversity and agroecology in the regulation of agriculture law is how this area of law is affected by the emergence of the new values of agrobiodiversity and agroecology.[4]

As regards agrobiodiversity, the value of species diversity affects agriculture law in terms of a new regulatory approach regarding agricultural activities, aimed at banning or discouraging some traditional farming systems which undermine biodiversity, such as the uses of herbicides, pesticides, and the unsustainable monocultures of industrial agriculture. Therefore, biodiversity becomes a paramount principle of agriculture law regulation, focusing on its protection.

It is an axiological process based on new regulation to promote some values, such as making agriculture sustainable for future generations. This value-oriented solution is not restricted to the traditional boundaries of agriculture law and implies changes in many other areas of private law, such as environmental protection and consumer information,[5] according to the trend described in Section 2.1.

For example, through mandatory labeling, consumers can be informed about how agricultural products were harvested and whether or not it was in compliance with the requirements of agrobiodiversity.[6]

Law systems will scarcely provide for a general definition of agrobiodiversity[7] or introduce a general regulation about agrobiodiversity.[8] It is more likely that a number of specific rules about farming techniques will be introduced, and consumer information about them may be mandatory too in order to create a regulation in favor of producers which opts for stronger protection of biodiversity.

As regards agroecology in legal terms, this is a broader concept, which considers the respect of ecological principles (not only biodiversity) in agricultural activities. Private law solutions to enhance an agroecology-oriented approach in agriculture law are similar to what has been considered with reference to agrobiodiversity: principally, this value-oriented approach will rely on environmental protection and consumer information.

2.3 *Agrobiodiversity and agroecology as part of private law*

This chapter shows that, in legal terms, agrobiodiversity and agroecology are values which can lead to changes not only in the field of agriculture law but also, in general terms, in private law relationships. A clear example of this is in the field of consumer law, such as when a new law prescribes mandatory labeling for some information related to agrobiodiversity and agroecology.

This means that consumers must have some information available before the product can go on sale, with reference to the degree of respect, on the part of the producer, to agrobiodiversity and agroecology principles: for example, to know whether or not a specific seed oil has been produced according to techniques that rule out intensive farming.

This legislative solution gives trade descriptions a new function, directly aimed to increase consumers' knowledge of the product and to enhance awareness of the environmental profiles of its production. As stated above, the function of this legislative approach[9] is not limited to agriculture law but includes general private law profiles in order to promote sustainable production techniques, by granting consumers preferences with regards to some products. In this way, a typical private law and consumer law prescription, such as mandatory labeling for some information, is not used for the general goal of enhancing consumers' knowledge, but for a specific issue. It becomes a spin for orienting the market toward sustainable agricultural products. It is a clear example of how traditional private law and consumer law structures can be used for new goals, such as for the protection of important values of our society in modern times.[10]

If this regulatory process leads to significant changes in private law, agrobiodiversity and agroecology will be considered as values which result in a comprehensive law reform and will therefore be counted as general principles of agriculture law. This process is under way, and, if legislators proceed accordingly, it will lead to a comprehensive law reform that will benefit both consumers and the environment.

2.4 Editors' note

In this chapter, Alberto Giulio Cianci explains the relationship between private law, agriculture law, agrobiodiversity, and agroecology. He outlines the fact that law systems do not provide a general definition of agrobiodiversity and agroecology, although they represent key issues in agriculture law. He also explains that, in the general system of private law, the relationship between consumer and producer has some specific interactions with reference to agrobiodiversity and agroecology—for example, an information issue, like mandatory labeling.

It is important to evaluate the position of agrobiodiversity and agroecology in the field of agriculture law and their private law-related issues in order to provide a better understanding of these concepts to the reader. This volume is a joint effort of several scholars, in addition to Alberto Cianci, whose papers aim to give the reader a comprehensive description of new trends related to agrobiodiversity and agroecology, as values able to influence a regulatory approach.

Gabriela Steier gives an example of how the new values of agroecology can influence law systems as a general trend which will affect international trade. Michael Blakeney focuses on a key issue in order to ensure food sustainability from an agroecology perspective that implies changes in trade regulation. Blakeney gives another example of how agrobiodiversity and agroecology can lead to changes in farming, based on traditions which have interesting intellectual property profiles.

Susannah Chapman and Paul J. Heald describe a crisis in agrobiodiversity that can be solved through a new regulatory approach aimed to protect diversity. Sumit Saurav and Wele Elangwe and Rosemary Agbor give two regional examples of this trend, involving corporate social responsibility profiles.

Ivan K. Mugabi focuses on a specific profile, oriented on the participation of disabled people in agroecology and agrobiodiversity.

Alexander Cherry gives a comprehensive analysis of olive oil production in Greece, outlining the importance of sustainability in new agricultural trends.

All these essays are strongly related to many areas of Private law—for example, trade regulation, intellectual property, disabled people. These areas are specifically mentioned, as they do not represent traditional profiles of agriculture law. Therefore, private law, not only agriculture law, can be considered as a link between all these essays, which it will be hoped will lead to debate about which trend of law reform can promote agroecology and agrobiodiversity as values for our society.

Even though Cianci's chapter provides only general links between agriculture law and private law, the links of private law protectionism with those areas of agriculture law that fail to protect agroecology and agrobiodiversity have made the connections abundantly clear. "BigFood" and "BigAg" have been served by private law protectionism on their path toward an oligarchy of the globe's food system. Nonetheless, looking back at the core links of agriculture law and private law, it becomes evident that private law may also serve agroecology and agrobiodiversity, poles apart from BigAg and BigFood's unsustainable practices.

The following chapter, written by Gabriela Steier, adds the concepts of the Right to Food and the UN's Sustainable Development Goals to this discourse. Together with the first two chapters, this third chapter takes the discussion of the fundamentals of legal protections of agroecology and agrobiodiversity full circle.

Notes

1. On the general principles and functions of food labeling with reference to EU law, see D. Lauterburg, *Food Law: Policy and Ethics* (London: Cavendish, 2001), 133 ff.; see also F. Parasecoli, *Knowing Where It Comes from* (Iowa City, IA: University of Iowa Press, 2017), 49 ff.
2. On transparency in food law, see N. Conte-Salinas, W. Wallau, The concept of transparency and openness in European food law, in G. Steier, K. Patel (eds), *International Food Law and Policy* (New York: Springer, 2016), 581 ff.
3. On the new functions of its definition, see A. Jannarelli, *Profili giuridici del sistema agro-alimentare tra ascesa e crisi della globalizzazione* (Bari: Cacucci, 2011), 91 ff.
4. On Agrobiodiversity law, see C.G. Gonzalez, Climate change, food security, and agrobiodiversity: toward a just, resilient, and sustainable food system, 22 *Fordham Environmental Law Review* 493 (2011); C. Ougamanam, Agrobiodiversity and food security: biotechnology and traditional agricultural practices at the periphery of international intellectual property regime complex, 2007 *Michigan State Law Review* 215 (2007); C. Fowler, Protecting farmer innovation: the convention on biological diversity and the question of origin, 41 *Jurimetrics* 477 (2000–2001).
5. With reference to food law, on consumer protection, see B. Atwood, K. Thompson, C. Willett, *Food Law*, 3rd ed. (Haywards Heath: Tottel, 2009), 139 ff.
6. On the ethical profiles of food safety, see C. Reimann, *Ernährungssicherung im Völkerrecht* (Stuttgart: Boorberg, 2000), 115 ff., 125 ff.; on the relationship between food safety and farmers' rights, see L. Winter, Cultivating farmers' rights: reconciling food security, indigenous agriculture, and TRIPS, 43 *Vanderbilt Journal of Transnational Law* 223 (2010).

7. On Agrobiodiversity as a juridical concept, see J. Santilli, *Agrobiodiversity and the Law* (Abingdon: Earthscan, 2012), 1 ff.; on its international law profiles, see M.A. Mekouar, Treaty agreed on agrobiodiversity: the International Treaty on Plant Genetic Resources for Food and Agriculture, 32 *Environmental Policy and Law* 20 (2002).
8. As a model or regulation for Peruvian Ley 27811—Régimen de protección de los conocimientos colectivos de los pueblos indígenas vinculados a los recursos biológicos (On protection of the collective knowledge of Indigenous people with reference to biological resources), see B. Venero Aguirre, La protección legal de los conocimientos tradicionales en el Perú, in P. Ferro, M. Ruiz (eds), *Apuntes sobre agrobiodiversidad* (Lima: Sociedad Peruana de Derecho Ambiental, 2005), 17 ff. On Agrobiodiversity as an international law concept, see A.P. Kameri-Bote, P. Cullet, Agrobiodiversity and international law: a conceptual framework, 11 *Journal of Environmental Law* 257 (1999).
9. On international law solutions to supervising states' regulatory approaches, see G. Westerveen, Towards a system for supervising states' compliance with the right to food, in P. Alston, K. Tomaševski (eds), *The Right to Food* (Utrecht: Martinus Nijhoff, 1984), 119 ff.; see also M.L. Maier, *Lebensmittelstandards und Handelsrecht im Verbund internationaler Regime* (Wiesbaden: Springer, 2017), 90 ff.; C. MacMaoláin, *Food Law* (Oxford: Hart), 2015, 51 ff.; N.D. Fortin, *Food Regulation* (Hoboken, NJ: Wiley, 2009), 637 ff.; C.I.P. Thomas, *In Food We Trust* (Lincoln, NE: University of Nebraska Press, 2014), 85 ff.
10. On a human rights dimension of international food law, see H. Schebesta, B. van der Meulen, M. van der Velde, International food law, in B. van der Meulen (ed.), *EU Food Law* (Wageningen: Wageningen Academic Publishers, 2014), 85 ff.

chapter three

Rights-based international agroecological law

Gabriela Steier*

Contents

3.1 Introduction

Plants are so useful
To me and to you.
Can you think of the ways?
I will name you a few.
...
The grains and the fruits
And the veggies you eat –
Why, they come from plants, too
And here's something neat ...

* A version of this chapter was first published as Small farmers cool the planet: The case for rights-based international agroecological law, *Groningen Journal of International Law* (2017).

Yes, plants serve us well,
And fill so many needs,
And flowering plants,
All started as seeds.

Bonnie Worth and Aristides Ruiz, *Oh Say Can You Seed?*

The foregoing rhyme is from a children's book. Such simple and beautiful words describing the usefulness of seeds and flowering plants that grow fruit resonate heavily with the overcomplicated reality of how such fruits reach consumers: the "farm-to-fork" path. No longer are idyllic and naive farm impressions a reality—and if they are, they seem expensive, for they internalize many of the costs that industrial farms and monocultures externalize, i.e., shift to the public, the planet, the future. Thus, the question is, what can we do to return to the idealized farms that we see in children's books more often than in reality? In other words, how can we protect the plants, animals, and wildlife that paint this wholesome picture for children? Why has the law not caught up with the protections that these delicate systems require? Children would ask: Why isn't anyone doing something to help?

In some ways, the wholesome pictures referred to previously are agroecology and agrobiodiversity. They depict a harmony of growing food and conserving the environment and its resources. Simply put, sustainability and resilience are elementary and intuitive. Sustainable food production should, therefore, be a core value of international agricultural trade, but it is not.[1] This chapter explores some of the challenges of achieving agroecology. One way of accomplishing this is a top-down approach, from treaties to nations and from nations to regions. Incorporating these goals into treaties touches on food trade.

The goal to incorporate sustainability into trade seems lofty and overly ambitious. It should not be, because this goal touches on the important rights-based equality between trading partners.[2] Its importance rests in the food trade, where developed countries often target developing countries in order to create a market for their surplus.[3] The underlying trade-distorting measures that developed countries use to ensure the functioning of their domestic food supply and the sales of surplus production lead to food dumping, a practice describing the surplus sales of overproducing Western countries to weaker developing markets.[4] Conversely, developing countries' markets are flooded with inexpensive commodity foods: often processed, input-intensive, high-calorie, and low nutrient-density snack foods. It follows that the agricultural market in developing countries is reoriented toward non-food crops, disturbing developing countries' agricultural exports,[5] because farmers no longer produce food for their local markets.

Even though this problem is oversimplified here, the resulting trade distortions create more universal problems, such as food insecurity, social unrest, unsustainable food production, environmentally harmful farming, and political uncertainty.[6] Scholars warn that "[c]urrent food insecurity is not caused by absolute food scarcity, but [is] the consequences of ineffective global food distribution, which is the result of distorted international trade,"[7] facilitated by Regional Trade Agreements (RTAs). RTAs are "deep integration partnerships between countries or regions with a major share of world trade and foreign direct investments."[8] These RTAs often fail to address the inequalities of trading partners and miss the important goal of trading governments to ensure food security. Some of the trade distortions could be addressed by combining food security and agroecology through a rights-based approach.

3.2 The rights-based approach to agroecology

The primacy of food for the survival of humans is a universal premise of this rights-based approach. It should be supported through food policies and legal frameworks.[9] One method of ensuring food security is through the Right to Food, defined by the United Nations (UN) as:

> the right of every individual, alone or in community with others, to have physical and economic access at all times to sufficient, adequate and culturally acceptable food that is produced and consumed sustainably, preserving access to food for future generations.[10]

As Professor Schutter from the University of Louvain and the former Special Rapporteur on the Right to Food notes, "The right to food can be summarized by reference to the requirements of availability, accessibility, adequacy and sustainability, all of which must be built into legal entitlements and secured through accountability mechanisms."[11] This chapter zooms in on the sustainability prong where the aforementioned legal entitlements and accountability mechanisms should be required elements of international trade agreements. For the reason that victims of food dumping need redressability for violations of their food security, whether in the past, present, or future, international agroecological law may help pave the way toward this rights-based approach by focusing on the aspect of sustainable food procurement.

Agriculture in tune with nature—so-called agroecology—has emerged as a more resilient and robust alternative to industrial food production.[12] This agroecological alternative should be central to the legal entitlements and accountability mechanisms in international trade agreements.[13] In the following sections, this chapter explores the necessity of incorporating agroecology as an important aspect of the Right to Food. Specifically, this chapter makes the case for a rights-based international agroecological legal framework that should guide RTAs to further the United Nation's goals on food security. The chapter then introduces agroecology as an emerged legal discipline and, subsequently, juxtaposes agricultural exceptionalism and the exceptionalism of agroecology. Next, the chapter continues to explore avenues through which agroecology could be integrated into international agricultural trade by awarding agroecology legal protection and examines some points of critique.

3.2.1 *Agroecology: an emerged discipline*

Agroecology essentially applies ecology to agriculture and has the ability to change the common vision of both agriculture and society.[14] Potentially capable of permeating various levels of society and environmental conservation, agroecology can be truly transformative for local economies and even international trade.[15] According to the World Bank—a stakeholder in the economies of developing countries—"Gross domestic product (GDP) growth in agriculture has been shown to be at least twice as effective in reducing poverty as growth originating in other sectors."[16] Agroecology, understood through the lens of the rights-based approach and environmental law, can help people achieve these goals.

Historically, agroecology links agriculture to both ecology and food systems. When agroecology first emerged as a discipline, it combined agronomy and ecology.[17] In its second phase, as Professor Monteduro from the University of Salento observes, "the interdisciplinary nature of agroecology extended to become inseparable from the problem of food."[18] This fundamental evolution integrated sociological, economic, political, historical, and scientific aspects into the field of agroecology.[19] As a result, agroecology now comprises organization, management, and development of agri-food systems, including production and consumption, thereby practically integrating agriculture into the concept of food systems. Now, in the third phase, agroecology has become "a fully-fledged transdisciplinary science," according to Monteduro.[20] By incorporating philosophical and bioethical sciences, and by "integrat[ing] … the theory of social systems applied to agricultural ecosystems, agroecotourism, social agriculture, urban and peri-urban agriculture, the rural landscape, the relations between rural communities and society, biotechnologies, [and] agroenergy,"[21] this interdisciplinary field responds to the conventional model of industrial agriculture with an eco-centric, culturally aware approach. All of these nuances should be factored into agricultural trade in order to introduce agroecological principles into RTAs, feasibly achieved by separating agriculture from economics.

Economically advantaged countries regulate and protect their agricultural systems more than poorer countries, sometimes at the expense of the latter. Protective mechanisms range from subsidies and tariffs to other barriers that allow richer countries to overproduce and dump their surplus on poorer countries.[22] It follows that existing "food policies and legal frameworks, as diverse, complex and globally scattered as they are, render it difficult to streamline food system regulation."[23] RTAs have consequently emerged as alternatives to this multilateralism in food systems. Complicating the complexity of these RTAs further, scholars observe that their "scale and scope ... have been eclipsed by the level of ambition of some new 'mega-regional' negotiations ... which have the potential to significantly reshape the global trade landscape."[24] At this level, the rights-based approach to agroecology should be factored into how RTAs may comply with global food systems. Especially because of the enormous potential of these RTAs to change food systems in developing nations,[25] the Right to Food and its underlying legal entitlements and accountability mechanisms must be considered.

This regulatory compatibility,[26] albeit fragmented and diverse around the globe, currently overlooks important aspects of sustainable food production and resilient food system regulation, whereby agricultural protectionism in international trade is facilitating food dumping in developing countries. As a result, developing countries experience nutrition shifts, causing a host of public health, food security, and also environmental problems. Schutter explains that:

> the increased reliance on food imports is a major cause of 'nutrition transition' in the developing world, by which nutritionists mean the shift to processed foods richer in salt, sugar and saturated fats—foods that have a long shelf life and are attractive to urban populations and younger generations, but are often less nutritious and less healthy.[27]

Moreover, food prices are highly volatile and foster social and political instability in many of the least developed countries.[28] Consequently, "small-scale farming was not viable under these conditions, many rural households [are] relegated to subsistence farming, surviving only by diversifying their incomes."[29] However, farmers can only diversify their incomes by applying agroecological methods, such as intercropping, and essentially moving away from the industrial model of conventional farming. Accomplishing this shift requires regulatory compliance and compatibility, which, in turn, is affected by RTAs on a higher governance level.

It is this diversification by way of agroecology on which this chapter focuses. If agroecology were protected as a branch of food security by way of the Right to Food, then agroecology as a legal subspecialty should, arguably, guide RTA negotiations and enactments. Although this idea is complex and daunting, this chapter asks: How could the implications of this agroecological twist on RTAs be justified?

3.2.2 *Agricultural exceptionalism and the exceptionalism of agroecology*

To date, RTAs set agriculture apart. Agriculture has been the most protected sector in international trade because governments seek to feed their people and provide jobs through the ever-needed food sector.[30] One possible justification is agricultural exceptionalism, "the use of legal exceptions to protect the agricultural industry."[31] This special status for agriculture as an industry "is evident throughout the law, with farmers protected from involuntary bankruptcy, exempted from many environmental regulations, and excepted from anti-trust restrictions."[32] Many exceptions for the agricultural sector occur in different countries, making exceptionalism a matter of international trade.[33] Using this special status of agriculture, however, could also support special considerations to protect the continued supply, i.e., the sustainability aspects. An economically removed aspect of food trade, the rights-based approach, for instance, could echo agricultural entitlements already resting on food's primacy. Thus, the successful interaction of such goals will eventually reconnect to development in a more uniform and equal manner, promising fairness in global food trade. The UN has already begun this process through the Sustainable Development Goals (SDGs).

The SDGs are a plan of action for people, planet, and prosperity, setting an aspirational agenda to transform the world by 2030 to eradicate poverty, an indispensable requirement for sustainable development.[34] The UN declares that the SDGs "are integrated and indivisible and balance the three dimensions of sustainable development: the economic, social and environmental."[35] Deutsche Gesellschaft für Internationale Zusammenarbeit [GIZ] (German Federal Ministry for Economic Cooperation and Development) special unit, "One World—No Hunger," a subdivision linking food security and sustainable agriculture, notes that "[s]everal SDGs touch the issue of conservation and sustainable use of agrobiodiversity"[36]—in other words, agroecology. Goals 2, 14, 15, and 16 in particular relate to agroecology within the food sovereignty framework:

> Goal 2: End hunger, achieve food security and improved nutrition and promote sustainable agriculture;
>
> Goal 14: Conserve and sustainably use the oceans, seas and marine resources for sustainable development;
>
> Goal 15: Protect, restore and promote sustainable use of terrestrial ecosystems, sustainably manage forests, combat desertification, and halt and reverse land degradation and halt biodiversity loss;
>
> Goal 16: Promote peaceful and inclusive societies for sustainable development, provide access to justice for all and build effective, accountable and inclusive institutions at all levels.[37]

These selected goals are some aspects of what agroecology can and should accomplish in a rights-based approach.[38] The second goal ties food sovereignty and food security together while promoting sustainable agriculture in line with agroecology. Similarly, Goal 14 emphasizes environmental conservation, while Goal 15 even adds biodiversity into the equation, again relating back to agrobiodiversity in food production and, thereby, agroecology. Notably, Goal 16 raises all of these issues into a societal and political domain, acknowledging the links between peace and sustainable development. Accomplishing Goal 13, "[t]ake urgent action to combat climate change and its impacts," is implied in the agroecology approach described here, but a detailed analysis is beyond the scope of this chapter. Nonetheless, the IPES report on the SDGs, albeit not making all the connections that this chapter makes, explicitly supports the agroecological principles.

From an environmental law perspective, agroecology and food sovereignty can be linked through the SDGs. Schutter observes that "[a]griculture is at a crossroads"[39] because "increasing food production to meet future needs, while necessary, is not sufficient"[40]—only sustainable methods of food production can ensure a continued supply of food for the growing population of the future. SDGs are one UN model to accomplish just that while supporting the case for international agricultural law. Similarly, the Aichi Biodiversity Target 13 of the Convention on Biological Diversity (CBD), Strategic Plan for Biodiversity 2011–2020, also supports agroecological goals and may benefit from strengthened agroecological advocacy in international trade. Target 13 provides:

> By 2020, the genetic diversity of cultivated plants and farmed and domesticated animals and of wild relatives, including other socio-economically as well as culturally valuable species, is maintained, and strategies have been developed and implemented for minimizing genetic erosion and safeguarding their genetic diversity.[41]

Thus, the SDGs and CBD illustrate how international goals may align with agroecology. As a counterpart to the forced and industry-dominated *agricultural* exceptionalism, *agroecological* exceptionalism lies in its nature by way of its harmonious coexistence with self-perpetuating biological processes. Notably,

> [a]gricultural biodiversity provides environmental services (soil, water, habitat, and pollinators) and supports the sustainability and resilience of agricultural systems; it can provide a diverse and

> nutritious diet, contribute to health, and support the maintenance of traditional knowledge and cultural identity.[42]

It follows that these traits of agroecology are certainly in line with the Right to Food and the principles of food sovereignty and food security embedded therein.

Investments are needed to promote a sustainable model of food production that will, in fact, feed the world. The UN Food and Agriculture Organization (FAO), however, estimates that "yearly investment in agriculture needs to rise by more than 50 percent," thus US$ 83 billion is required annually to meet the SDGs by 2030.[43] These goals include concerns about how long-term gains in food production can be reconciled with environmental and resource conservation and ecosystem protection. Simply put, agroecology's goals are evidently beyond economic gains because "short-term [economic] gains will be offset by long-term losses if it leads to further degradation of ecosystems, threatening [the] future ability to maintain current levels of production."[44] Recognizing these trends, Schutter observes, governments are paying more attention to agriculture and agri-food companies have increased their investment in the long-term viability of supplies, with foreign average annual investments rising from US$ 600 million in the 1990s to US$ 3 billion in 2005–2007.[45] Governments may, consequently, be receptive to consider the rights-based agroecological approach in RTA negotiations. Especially, as Schutter explains, conventional agriculture, while supposedly producing a sufficient amount of food, will fail to produce nutritionally adequate, environmentally sustainable, and thereby, continuously available food. In other words, the current industrial model of food production will not feed the world in the future.

For instance, through the practice of agroecology, farmers can diversify their incomes by diversifying the crops they grow and variegating the methods they utilize to cultivate their land. Some governments already concede that

> [a] rich diversity of native plant varieties and locally adapted animal breeds contributes to strengthening these farmers' and herders' resilience in the face of difficult climatic conditions and marginal locations, e.g. in arid or upland regions. Traditional crops and livestock breeds can be utilized with minimum agricultural input, have quality characteristics that correspond to local needs and also often play an important role in the culture of the rural population.[46]

The diversification of agriculture with the goal to promote sustainable and climate-change-resilient farming practices is the backbone of this system—and this system is in dire need of legal protection.

3.3 Legal protection for agroecology's infrastructure

Implementing the rights-based approach to agroecology means strengthening the infrastructure of laws, treaties, and regulations that foster an agroecology-friendly trading environment. Promoting agroecological research and separating biocultural goals from economic drivers may be one method to change the legal landscape. For instance, Monteduro links the SDGs to agroecology by way of biocultural diversity, which he considers key.[47] He explains that, "[o]n the one hand, bio-cultural diversity is shared between food sovereignty and agroecology … on the other, many SDGs implicitly or explicitly refer to bio-cultural diversity."[48] These observations align with those of Belgian scholars, who found that agroecological research can help to "develo[p] tools and methods for better understanding ecosystem services, evaluating their importance, optimizing natural processes, developing socioeconomic systems for paying their production, and integrating ecosystem services in the intrinsic mechanisms of the society of tomorrow."[49]

For example, in Belgium, three-quarters of plant production is used for animal feeding, fueling meat consumption 40 percent higher than what is recommend by nutritionists.[50] From an agroecological standpoint, this "means that part of the land could be devoted to crops other than annual crops for animal feeding, which creates opportunities for diversification and for a larger share of (permanent) grasslands in the agricultural area."[51] This diversification may halt overproduction and solve part of the dumping problem mentioned previously.

Agroecology, as a cross-disciplinary field, has potential to reach further to improve nutritional outcomes by diversifying diets through a more varied food supply. In the USA, an internal Food and Drug Administration (FDA) report titled, "The Nutrition Review Project," which examined policies intended to make Americans healthier, concluded that a complete reset is needed.[52] Award-winning journalist Naomi Klein observes that the current debate about agriculture contrasts industrial agriculture's higher yields and local or organic farming's lower chemical inputs and shorter supply chains.[53] She considers agroecology "a less understood practice in which small-scale farmers use sustainable methods based on a combination of modern science and local knowledge."[54] A diversification of agrobiodiversity and the proliferation of agroecological practice could contribute to the systematic overhauls that are necessary, thereby returning local and high-quality nutrient-dense foods to society that the industrialization of agriculture has artificially removed from our plates.[55] After all, diversifying the food supply may help promote nutritional adequacy.

The implications of incorporating the rights-based approach, however, has been criticized for "impoverishing political discourse" because "[t]he absoluteness of human rights may promote unrealistic expectations, heighten social conflict and inhibit dialogue that might otherwise lead towards consensus, accommodation or at least discovery of common ground"[56] —this is an argument from industry lobbyists. In 2008, the year the previous US Farm Bill passed, BigAg mega-corporation Monsanto, for instance, spent $8.8 million in lobbying expenditures, $8 million in lobbying expenses in 2010, another $6.37 million in 2011, and nearly $6 million more in 2012.[57] The Union of Concerned Scientists, a national non-profit organization, reports that Monsanto's reported lobbying in early 2011 succeeded at creating a so-called "modern agriculture" caucus in Congress,[58] which further evidences that the dialogue leaders in the agricultural policy are the stakeholders who could benefit most from RTAs. Monsanto is an example of a large international stakeholder that seeks to stifle honest dialogues advocating for agroecology because diversification would harm Monsanto's bottom line. Multinational companies like Monsanto, Dow, Syngenta, and Bayer have stakes in RTAs and hold the global system in a gridlock. This chapter seeks to inspire the discourse for a legal framework in favor of agroecology that sets RTAs free from the unilateral economic orientation that, as previously described, leads to industrial agriculture, the consequential overproduction, dumping, and weakening of developing countries' food security.

Sustainability, albeit a vague term, can better restore food security in an agroecological framework. Professor Laurie Ristino, Director of the Center for Agriculture and Food Systems at the Vermont Law School, writes that

> nearly every step in the sustainable food chain requires law to support it. The policy work done to date is a fine start, but without the legal infrastructure to undergird policy, it will have limited traction in our rule of law society.[59]

This observation expands through international agricultural trade, where diversification should be valued over uniformization, as is currently occurring. Thus, according to the IPES report, agroecological law can help to turn lock-ins into points for change, as illustrated by Figure 3.1.[60] Programs that support these paradigm shifts toward agroecology in international food trade include the 2005 Millennium Ecosystem Assessment, 2009 International Assessment of Agricultural Knowledge, Science and

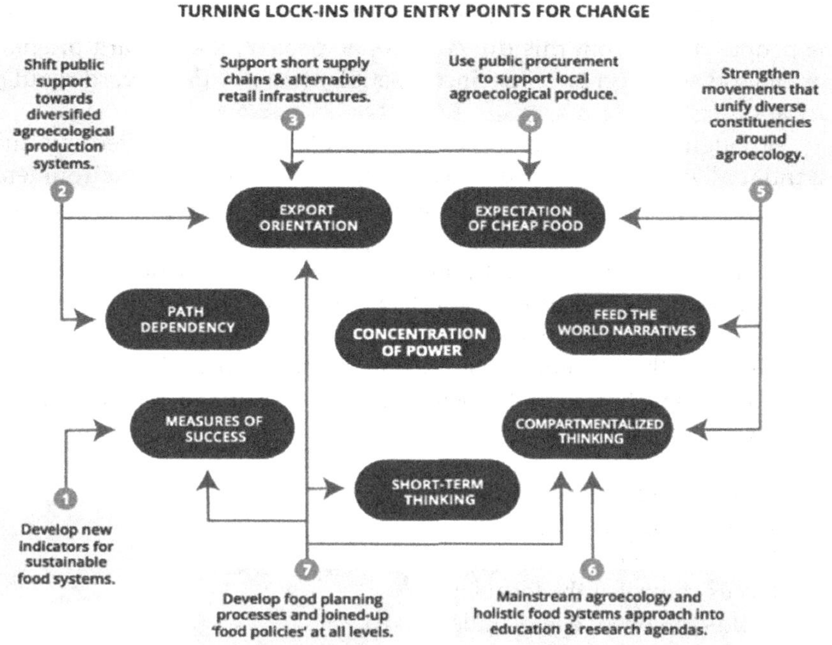

Figure 3.1 Turning lock-ins into entry points for change. The lock-ins (black) can be unlocked by agroecological legal strategies (grey), thereby creating opportunities for a transition to diversified agroecological systems. See Endnote 60 for image source.

Technology for Development (IAASTD) with over 400 cited supporting studies, FAO's regional agroecology meetings and its training courses to build agroecology into its Farmer Field School systems in 2015 and 2016, and FAO and UNEP's Sustainable Food Systems Programme (SFSP) as part of the UN 10 Year framework programme on sustainable consumption and production.[61]

3.4 *Disconnecting critiques*

The rights-based legal protection for agroecology described herein is controversial. One of the main objections to grouping agroecology, food sovereignty, and environmental law (through the SDGs) into the rights-based approach is their mismatched directionality. Simply put, food sovereignty, the grassroots movement initiated by La Via Campesina, takes a bottom-up approach, while both agroecology and the SDGs are top-down approaches whereby scientists evaluate and abstract or governments legislate and regulate. Framing this another way, food sovereignty is a principle *by* the people, while agroecology and environmental law are *for* the people. Thus, from this directional perspective, the upward orientation of food sovereignty seems incompatible with any downward regulatory approach.

Although this mismatch seems logical, it is discredited by the essential alignment of all three parts: that of agroecology, environmental law, and food sovereignty. The mere fact that the former UN Special Rapporteur on the Right to Food put agroecology within the context of human rights, specifically the Right to Food, already provides an authoritative link between the three parts. The FAO and Professor Schutter also provide a conclusive connection between food sovereignty and agroecology by linking the concepts in various publications, including the Nyéleni Declaration, which essentially describes agroecological goals under the food sovereignty umbrella:[62]

1. Focuses on food for people
2. Values food providers
3. Localizes food systems
4. Places control at the local level
5. Builds knowledge and skills
6. Works with nature[63]

Another objection originates from the hegemonic monism of industrial agriculture, taking the position that "only a few changes are necessary to apply those tools to environmental problems."[64] As Figure 3.1 shows, one such change consists of decoupling agroecology and economic goals in food trade regulation.[65] The objectivist (but not necessarily objective) view, links the "costs (loss of nutrients and biodiversity and environmental degradation) and benefits (production, generation of wealth, and maintaining the environment) of agriculture,"[66] but erroneously focuses on "the resources that enter and abandon the agricultural systems [which] are seen as finite capital measured in monetary or physical units"[67] overlooking non-quantifiable aspects, such as biodiversity losses, soil depletion, and environmental degradation. A truly objective approach would also consider that "the capital that enters and exits agricultural systems is not measured only in physical units, but also includes cultural knowledge, human experience [and] potential for technological development."[68] Alarmingly,

> statutes that are putatively designed to protect the environment are often more honestly described as programs for boosting commodity prices and farm incomes by restricting output. For example, the Soil Conservation Act of 1936 described wheat

> as a 'soil-eroding' crop and soybeans as a 'soil-conserving' crop, in apparent defiance of agronomy but conveniently in accord with the income-support provisions of the invalidated Agricultural Adjustment Act of 1933.[69]

Another example is the Conservation Reserve Program (CRP), which "has failed to produc[e] benefits sufficient to cover its costs."[70] To be sure, neither the environmental benefits nor the fiscal costs of the CRP can be quantified with an absolute degree of confidence[71] and the rights-based approach toward agroecology should prohibit the abuse of subsidies, as indirect bribes to farmers. Thus, "[i]f indeed farmers are 'stewards' of the land, they are among the most richly bribed guardians of environmental integrity."[72]

Instead of shutting out the rights-based approach, "agroecological thought should open up to epistemological pluralism for production of agricultural knowledge."[73] This concept "goes beyond overcoming the disciplinary compartmentalization characteristic of conventional science, since it questions the hegemonic belief of the superiority of scientific practice."[74] Admitting that, there is no consensus in the literature on "whether said incorporation is instrumental (e.g. the use of cropping techniques or associations among species) or epistemological (i.e., the articulation of scientific discourse with forms of non-scientific knowledge, such as peasant, indigenous or afro knowledge)."[75] There is, however, a general tendency[76] that considers "agroecology … [to be] a scientific discipline that integrates different disciplines"[77] and practices.[78] The epistemological pluralism and interdisciplinary nature of agroecology within various climatic, cultural, and economic contexts are the ultimate justification that neither conceptual depth nor discussion is necessary or useful to advance the field. Agroecology is, after all, rooted in biodiversity, evolution, and adaptability—virtues tackling the core problems of the current food system, where industrial agriculture has failed.

Groundswell International, a non-profit organization with a mission to strengthen rural communities in order to build healthy farming and food systems from the ground up, reports that:

> while over the past century industrial agriculture led to increases in global staple foods production through the use of pesticides and fertilizers, it has failed to eradicate world hunger and instead contributed to increased poverty rates. It has shifted production from multitudes of farmers to few producers and reduced soil quality while bearing a heavy burden on our planet.[79]

Groundswell International cites statistics from a 2016 International Panel of Experts on Sustainable Food Systems (IPES) report:

- Crop yields failed to improve, stagnated or collapsed in 24–39% of the world's maize, rice, wheat, and soybean production zones over recent decades.
- Large-scale producers deprive farmers of their food sovereignty.
- Farmers constitute 50% of the world's poor.
- Global food systems account for one-third of greenhouse gas emissions and [are] a primary polluter of water sources and depleter of water tables and forests.[80]

In fact, IPES experts add that the "feedback loops or 'lock-ins' built into the industrial agricultural system ... keep us bound to it,"[81] and industrial agriculture "leads systematically to negative outcomes and vulnerabilities."[82]

3.5 Conclusion

Agroecology and food sovereignty are vital for the functioning of food systems and should be legally protected through the Right to Food in international trade. As an emerged legal discipline, agroecological protection severed from economic goals and in line with the SDGs should be at the forefront of RTA negotiations. Sustainable development, climate change resilience, and international trade, connected through food and agriculture, may point the way forward. Protecting these interests may require a streamlined approach to research and advocacy for these interrelated concepts on a broad global level. As this chapter describes, small farmers using agroecology may cool the planet and feed the world.

3.6 Editors' note

This chapter follows the first chapter by Blakeney on local agricultural knowledge and food safety and then Cianci's chapter on agricultural law and private law. Here, Gabriela Steier provides for a comprehensive rights-based approach to agroecology from the perspective of the law. She considers agroecology as a key issue toward food sovereignty, outlining the framework of the required legal protection for agroecology. This chapter focuses on which legal tools already exist to protect agrobiodiversity and agroecology, complementing the first two chapters in this volume.

Steier's chapter summarizes how agroecology and food sovereignty are vital for the functioning of food systems and should be legally protected through the Right to Food in international trade. As an emerged legal discipline, she posits, agroecological protection severed from economic goals and in line with the SDGs should be at the forefront of RTA negotiations. Trade distortions create more-universal problems, such as food insecurity, social unrest, unsustainable food production, environmentally harmful farming, and political uncertainty. Some trade distortions could be addressed by combining food security and agroecology through a rights-based approach. For the reason that victims of food dumping need redressability for violations of their food security, whether in the past, present, or future, international agroecological law may help to pave the way toward this rights-based approach by focusing on the aspect of sustainable food procurement.

In describing such processes, Steier evaluates the "dilemma" between a rights-based approach and a policy-based approach, citing some positions of industry lobbyists which support the second solution and the associated risks. These positions do not reflect the real relationship between law and policy, where the main principle in all fundamental rights systems is that they are not subject to political choices. If a system of fundamental rights is part of a modern legal approach to agroecology, policy choices in regulating industry business must be enacted only if they respect fundamental rights.[83] This key issue in Steier's chapter, which provides a comprehensive review of innovative legal solutions for regulating agroecology from the perspective of international law up to national law systems, may thus pave the way for a better harmony of law and policy to further agroecological goals and to protect agrobiodiversity through the law.

Notes

1. See generally UN, The role of the international economy, in *Our Common Future: Report of the World Commission on Environment and Development* (A/42/427) (1987), https://digitallibrary.un.org/record/139811.
2. Ibid.
3. Ibid.
4. Ibid.
5. See generally Oxfam Briefing Paper, *Dumping on the World: How EU Sugar Policies Hurt Poor Countries* (March 2004), www.oxfam.org/sites/www.oxfam.org/files/bp61_sugar_dumping_0.pdf.
6. Ibid.
7. Uneven food distribution and distorted agricultural trade: An overlooked factor, in Ying Chen, *Trade, Food Security, and Human Rights: The Rules for International Trade in Agricultural Products and the Evolving World Food Crisis* (London: Routledge, 2014), at 73.
8. Mariagrazia Alabrese et al., *AgLaw Colloquium* (2016) (adapted from the call for papers).
9. Michael Roberts in Chen, *Trade, Food Security, and Human Rights*, at i.
10. Olivier De Schutter, *Final Report: The Transformative Potential of the Right to Food* (A/HRC/25/57 24) (January 2014), at 3 (citing Committee on Economic,

Social and Cultural Rights, General Comment No. 12 on the right to adequate food, paras. 6 and 7).

11. Ibid.
12. Ibid.
13. Union of Concerned Scientists, *Counting on Agroecology: Why We Should Invest More in the Transition to Sustainable Agriculture* (2015), www.ucsusa.org/food-agriculture/advance-sustainable-agriculture/counting-on-agroecology#.V3uBD1cw1SU.
14. Fabio Caporali, History and development of agroecology and theory of agroecosystems, in Massimo Monteduro et al. (eds), *Law and Agroecology* (Berlin: Springer, 2015), at 3.
15. For the purpose of this chapter, agriculture is "[a] linked, dynamic social–ecological system based on the extraction of biological products and services from an ecosystem, innovated and managed by people ... encompass[ing] all stages of production, processing, distribution, marketing, retail, consumption and waste disposal." Caporali, History and development of agroecology, at 5.
16. FAO, *FAO's Role in Investment in Agriculture*, https://web.archive.org/web/20160221094830/http://www.fao.org/investment-in-agriculture/en/ (last accessed June 30, 2016), citing World Bank, *World Development Report 2008: Agriculture for Development* (2008).
17. Massimo Monteduro, Environmental law and agroecology: Transdisciplinary approach to public ecosystem services as a new challenge for environmental legal doctrine, *European Energy and Environmental Law Review* (February 2013), at 2–11, 4.
18. Ibid.
19. Ibid.
20. Ibid.
21. Ibid.
22. Monteduro et al., Law and agroecology.
23. Rory Freeman and Jim Rossi, Agency coordination in shared regulatory space, 125 *Harvard Law Review* (2012) 1131, 1133.
24. Alabrese et al., *AgLaw Colloquium*. Examples of RTAs include: TTIP, TPP, RCEP, CFTA.
25. UN, the role of the international economy.
26. Alabrese et al., *AgLaw Colloquium*.
27. Schutter, Final Report, at 7.
28. Ibid.
29. Ibid.
30. Chen, *Trade, Food Security*, at 74.
31. Susan A. Schneider, A reconsideration of agricultural law: A call for the law of food, farming, and sustainability, 34 *William & Mary Environmental Law and Policy Review* (2010) 935.
32. Ibid., at 936.
33. Ibid.
34. UN, *Transforming Our World: The 2030 Agenda for Sustainable Development*, https://sustainabledevelopment.un.org/post2015/transformingourworld.
35. Ibid.
36. GIZ, *Agrobiodiversity: The Key to Food Security, Climate Adaptation and Resilience*, https://web.archive.org/web/20171110114228/www.giz.de/

fachexpertise/downloads/giz2015-en-agrobiodiversity-factsheet-collection-incl-mappe.pdf, at 2.
37. UN, *Take Urgent Action to Combat Climate Change and Its Impacts* (A/RES/70/1), www.un.org/ga/search/view_doc.asp?symbol=A/RES/70/1&Lang=E.
38. UN, The role of the international economy.
39. Schutter, *Agroecology and the Right to Food* (2010) (A/HRC/16/49), at 3.
40. Ibid.
41. Colin Khoury, Are we getting anywhere?, CIAT (International Center for Tropical Agriculture) (June 2, 2016), https://blog.ciat.cgiar.org/are-we-getting-anywhere/.
42. GIZ, *Agrobiodiversity*, at 4–5.
43. FAO, *Sustainable Development Goals*, https://sustainabledevelopment.un.org/sdgs. See also FAO, *Foreign Investment in Agriculture*, www.fao.org/economic/est/issues/investments/en/#.XMqu9mhKiHt.
44. Schutter, *Agroecology and the Right to Food*, at 3.
45. Ibid.
46. GIZ, *Agrobiodiversity*, at 4–5.
47. Email correspondence with Prof. Massimo Monteduro (July 7, 2016) (on file with author).
48. Ibid.
49. A. Peeters, N. Dendoncker, and S. Jacobs, Enhancing ecosystem services in Belgian agriculture through agroecology: A vision for farming with a future, in S. Jacobs, N. Dendoncker, and H. Keune (eds), *Ecosystem Services: Global Issues, Local Practices* (Boston, MA: Elsevier, 2013), at 287 (internal citations omitted).
50. Ibid., at 294–295 (internal citations omitted).
51. Ibid., at 295 (internal citations omitted).
52. Helena Bottemiller Evich, FDA memo urges reset of nutrition goals, *Politico* (August 10, 2016), www.politico.com/tipsheets/morning-agriculture/2016/08/fda-memo-urges-reset-of-nutrition-goals-jbs-transfers-parent-company-to-ireland-oj-keeps-declining-215809#ixzz4H0ujc6uY.
53. N. Klein, *This Changes Everything: Capitalism v. the Climate* (New York: Simon & Schuster, 2014), at Kindle location 2466.
54. Ibid.
55. Peeters et al., Enhancing ecosystem services, at 287 (internal citations omitted).
56. See generally Priscilla Claeys, *Human Rights and the Food Sovereignty Movement: Reclaiming Control* (London: Routledge, 2015).
57. Union of Concerned Scientists, *Lobbying and Advertising*, https://web.archive.org/web/20180307061110/https://www.ucsusa.org/food_and_agriculture/our-failing-food-system/genetic-engineering/lobbying-and-advertising.html.
58. Ibid.
59. Laurie Ristino, Back to the new: Millennials and the sustainable food movement, 15 *Vermont Journal of Environmental Law* (2013) 1, 22.
60. Figure 3.1 is taken from International Panel of Experts on Sustainable Food Systems (IPES), *From Uniformity to Diversity: A Paradigm Shift from Industrial Agriculture to Diversified Agroecological Systems*, June 2016, ipes-food.org/images/Reports/UniformityToDiversity_FullReport.pdf (accessed November 27, 2016).

61. Ibid.
62. Nyéléni, *Declaration of the Forum for Food Sovereignty* (February 2007), nyeleni.org/spip.php?article290 (accessed November 27, 2016).
63. See www.permaculture.org.uk/sites/default/files/page/document/Nyeleni_Food_Sovereignty.pdf.
64. Luis Fernando Gómez, Leonardo Ríos-Osorio, and Maria Luisa Eschenhagen, Epistemological bases of agroecology, 49 *Agrociencia* (2015), 679–688, 679.
65. Jim Chen, Get green or get out: Decoupling environmental from economic objectives in agricultural regulation, 48 *Oklahoma Law Review* (1995) 333, 3433.
66. Gómez, et al., Epistemological bases of agroecology, at 686.
67. Ibid. (internal citations omitted).
68. Ibid.
69. Chen, Get green or get out, at 343.
70. Ibid., 344.
71. Ibid.
72. Ibid.
73. Gómez et al., Epistemological bases of agroecology, at 681 (internal citations omitted).
74. Ibid.
75. Ibid.
76. Ibid., at 687 (internal citations omitted).
77. Ibid., at 682.
78. Ibid.
79. Groundswell International, *New IPES Report Advocates for Agroecology* (June 22, 2016), www.groundswellinternational.org/agroecology/new-ipes-report-advocates-for-agroecology/.
80. International Panel of Experts on Sustainable Food Systems (IPES), *From Uniformity to Diversity: A Paradigm Shift from Industrial Agriculture to Diversified Agroecological Systems* (June 2016), https://web.archive.org/web/20170302190309/http://www.ipes-food.org/images/Reports/UniformityToDiversity_FullReport.pdf (accessed November 27, 2016).
81. Groundswell International, *New IPES Report.*
82. IPES, *From Uniformity to Diversity*, at 3.
83. On the relationship between fundamental rights and agroecology, see M.A. Mekouar, Food security and environmental sustainability: Grounding the right to food on agroecology, 44 *Environmental Policy & Law* (2014).

part two

Specific challenges for agrobiodiversity and agroecology

chapter four

Regulatory options for food waste minimization

Michael Blakeney

Contents

4.1 Introduction

The loss or waste of consumable food is an enormous global problem. According to a report by the Food and Agriculture Organization of the United Nations (FAO), 1.3 billion tons of foodstuffs, around a third of all global food production, is lost or wasted every year.[1] The UN Secretary-General's High-Level Panel on Global Sustainability estimated that the amount of food wasted by consumers in high-income countries alone is roughly equal to the entire food production of sub-Saharan Africa.[2]

Food loss and waste (FLW) also has significant environmental impacts. The FAO has estimated that 3.49 billion tons of carbon dioxide are generated by food lost or wasted along its supply chain.[3] That adds up to approximately 14 percent of the world's CO_2 emissions.[4] Furthermore, food left to rot in landfills impacts the biodiversity around the landfill and pollutes waterways and groundwater.[5] The water used for irrigation to produce the wasted food could have met the domestic water needs of an estimated 9 billion people.[6]

FLW occurs at all stages along the supply chain. In developing countries, an estimated two-thirds of such losses occur at the post-harvest and processing stages due to inadequate agricultural practices and infrastructure for storage, processing, and transport.[7] In developed countries, most FLW occurs at the consumption level, driven by behavioral factors such as impulse or bulk buying and poor planning.[8]

The economic consequences of food waste are generally felt in the domains of household expenditures, raw material and agricultural expenditures, savings from avoidable food waste, costs for waste treatment, health costs from over-eating, and environmental costs attributable to food waste disposal.[9]

This chapter considers various regulatory options to minimize the generation of FLW. For this, it is necessary to understand the causes of FLW at each level of the supply chain and to acknowledge that the drivers of FLW will differ between developed and developing countries.

4.2 *Terminology and definitions*

A threshold issue in the consideration of legislative options for the minimization of FLW is the definition of the terms central to its regulation. These include: "food," "food loss," "food waste," and "food supply chain."

The FAO's Codex Alimentarius defines "food" as "any substance, whether processed, semi-processed or raw, that is intended for human consumption," including drinks, chewing gum, and any substance which has been used in the manufacture, preparation, or treatment of "food," but excluding cosmetics, tobacco, or substances used only as drugs.

The US Food Drug and Cosmetic Act (21 US Code § 321) uses a slightly different wording, defining "food" as "articles used for food or drink for man or other animals, chewing gum, and articles used for components of any such article." The European Food Safety Authority, which seeks to establish general principles and requirements of food law,[10] has a more detailed definition. In Article 2, European Regulation (EC) No. 178/2002, "food" is defined for the purposes of the regulation as "any substance or product, whether processed, partially processed or unprocessed, intended to be, or reasonably expected to be ingested by humans." This includes "drink, chewing gum and any substance, including water, intentionally incorporated into the food during its manufacture, preparation or treatment," but excludes:

(a) feed
(b) live animals unless they are prepared for placing on the market for human consumption

(c) plants prior to harvesting
(d) medicinal products[11]
(e) cosmetics[12]
(f) tobacco and tobacco products[13]
(g) narcotic or psychotropic substances within the meaning of the United Nations Single Convention on Narcotic Drugs, 1961, and the United Nations Convention on Psychotropic Substances, 1971
(h) residues and contaminants[14]

Article 3 of the European Regulation (EC) No. 178/2002 defines that, for the purposes of the Regulation, "food law" means

> the laws, regulations and administrative provisions governing food in general, and food safety in particular, whether at the community or national level; it covers any stage of production, processing and distribution of food, and also of feed produced for, or fed to, food-producing animals.[15]

According to the FAO, FLW refers to "the edible parts of plants and animals produced for human consumption but are not ultimately consumed by people."[16] In particular, *food losses* refer to quantities of food that are lost along the food supply chain and do not reach the ultimate consumer. In contrast, *food waste* is defined as referring to food that reaches the ultimate consumers in the desired quality but is not consumed and is instead discarded.[17]

Food waste is thus recognized as distinct from food loss, with different factors generating each one, meaning that different policies are required to deal with each phenomenon. Another distinction can be drawn between food losses and food waste, with losses as the result of "unintentional" events and waste as occurring more through conscious action or "negligence."[18] Defining food waste was a central feature of the recently concluded EC-funded FUSIONS project.[19] Their definition of food waste was:

> any food, and inedible parts of food, removed from the food supply chain to be recovered or disposed (including composted, crops ploughed in/not harvested, anaerobic digestion, bio-energy production, co-generation, incineration, disposal to sewer, landfill or discarded to sea).[20]

Since 2013, the World Resources Institute (WRI), in collaboration with a number of international partners,[21] has been developing a global Food Loss and Waste Accounting and Reporting Standard for quantifying food and/or associated inedible parts removed from the food supply chain.[22] The Protocol seeks to propose an internationally accepted accounting and reporting standard and associated tools, and to promote their adoption so entities are better informed and motivated to take appropriate steps to minimize FLW. It is useful to consider the definitions used in the Protocol since they are equally useful in formulating baseline definitions for legislative regulation of FLW. The glossary to the Protocol defines FLW as "Food and/or associated inedible parts removed from the food supply chain."[23] For the purposes of the Glossary the definition of "food" is modeled on that of the Codex Alimentarius Commission as "any substance—whether processed, semi-processed, or raw—that is intended for human consumption and includes drinks and any substance that has been used in the manufacture, preparation, or treatment of 'food.'" It includes material that has spoiled and is no longer fit for human consumption, but excludes cosmetics, tobacco, or substances used only as drugs, as well as processing agents used along the food supply chain— for example, water to clean or cook raw materials in factories or at home.

4.3 *Drivers of FLW*

A 2011 study by The Swedish Institute for Food and Biotechnology (SIK), completed for the FAO,[24] identified the causes of food losses and waste in low-income countries as "mainly connected to financial, managerial and technical limitations in harvesting techniques, storage and cooling facilities in difficult climatic conditions, infrastructure, and packaging and marketing systems." In medium- or high-income countries, however, the causes of food losses and waste "mainly relate to consumer behaviour or a lack of coordination between different actors in the supply chain."[25] Thus, farmer–buyer sales agreements in those countries were identified as a source of large quantities of farm crops being wasted—for example, by the imposition of quality standards, which reject food items not perfect in shape or appearance.

At the consumer level, in medium- to high-income countries, insufficient purchase planning and expiring "best-before dates" were also identified as the causes of large amounts of waste, in combination with a careless attitude of consumers who can afford to waste food.[26] In short, in those countries, the main drivers of food waste can be summarized as: exacting quality standards, excessive management regulation, poor environmental conditions during display, and consumer behavior.[27]

Despite the veracity of these generalizations based upon income categories, the circumstances under which food losses and waste occur are strongly dependent on the specific food- and waste-related conditions of each country, since everyone has its own unique production, processing, distribution, and consumption practices.

Regulation of food waste in a country will be more effective if it addresses the drivers of food loss and waste in that country. Regulation is probably the most effective when combined with educational programs to raise awareness of food loss and waste among food industries, retailers, and consumers.

4.4 Regulatory examples

Until the last 20 years of the twentieth century, regulation in this sector typically emanated from national legislatures, sometimes in order to address obligations imposed by international agreements to which a state had subscribed. In recent years, however, industry self-regulation, defined as "a regulatory process whereby an industry-level organization sets rules and standards relating to the conduct of firms in the industry,"[28] has become increasingly dominant.[29] Such self-imposed regulations, however, are not equivalent to those enforced by governments. In contrast with government-imposed laws, self-regulation is voluntary and typically framed as corporate social responsibility with public welfare as its central feature. Corporate self-regulation can be advantageous to state governments because of its flexibility, the voluntary consent of the regulated, and the conservation of government resources. Some important risks to the public sector, however, may arise from weak standards or ineffective enforcement. In some jurisdictions, these risks can be ameliorated by the overlapping of private and public regulation in the form of government scrutiny of self-regulation standards. For example, in Australia, the self-regulatory codes are subject to the scrutiny of the competition law authority to ensure the conferral of public benefit.[30]

The primary motivating factor for the development of self-imposed regulation has been the threat of legislative intervention because of social concerns.[31] Corporate self-regulation was pioneered in industries that provoke environmental concerns,[32] such as mining[33] and forestry,[34] although food and food-related industries have also become an important field for industry self-regulation. Some of these rules have been adopted to regulate the industry practices of marketing food to children and young people,[35] the promotion of unhealthy foods to schoolchildren,[36] including the supply of sugar-sweetened beverages[37] and television advertising,[38] as well as the advertising industry's practices in relation to alcohol.[39]

The representation of the public interest in industry self-regulation depends in large part upon the strength of the tradition of corporate social responsibility in the country concerned.

Different nations around the world have different traditions of regulation. In relation to food waste minimization, as explained in depth next, the United Kingdom (UK) has an effective self-regulation regime, underpinned by government funding. The government of the USA, on the other hand, has mostly maintained a laissez-faire policy in relation to the marketing of junk food to children, leaving the food industry largely to police itself, with modest guidance from federal regulators and none from consumer advocates.

In countries without a significant self-regulatory tradition, or where the issue of food waste minimization is considered to be too important to be left to private regulation, the state has intervened with direct legislation. A few examples from France, Italy, Germany, the Netherlands, and Japan are also described next.

4.4.1 Voluntary regulation: the UK

The key ingredient of the voluntary FLW minimization regime in the UK is the Courtauld 2025 commitment, an initiative launched and managed by WRAP (Waste and Resources Action Programme), a registered UK charity working with businesses, individuals, and communities to promote the reduction of waste and the development of sustainable practices. It was established in 2000 as a company limited by guarantee, receiving funding from the Department for Environment, Food and Rural Affairs, the Northern Ireland Executive, Zero Waste Scotland, the Welsh Government, and the European Union. WRAP has launched a number of initiatives, including "Recycle Now" and "Love Food, Hate Waste," with the aim of assisting businesses, local authorities, community groups, and individuals to reduce food waste. In 2005, it launched the Courtauld Commitment, a voluntary agreement signed by major UK supermarkets with the purpose of reducing waste across the UK grocery sector, renewed in 2010 and again in 2015.

Courtauld 2025, the fourth iteration of the agreement,[40] comprises a ten-year commitment by its signatory organizations to identify priorities, develop solutions, and implement what are determined to be best practices, all across the UK. The shared objective is to cut the resources needed to provide food and drink by one-fifth in ten years.[41] At the launch of the agreement, its signatories included all the major UK grocery retailers, together representing over 93 percent of the 2016 UK food retail market.[42] Commitment targets are estimated to deliver savings worth £20 billion to the UK economy and to satisfy its obligations under the UN's Sustainable Development Goal 12.3 to halve household and retail waste.[43]

WRAP's responsibility is to work directly with stakeholders to support actions in Courtauld 2015 under four main areas:

- Embedding sustainable principles and practices into the design, buying and sourcing of food
- Optimising resource efficiency throughout entire supply chains to help produce more goods using less resources

- Influencing behaviours around consumption and reduce waste in the home
- Finding innovative ways to make the best use of surplus and waste food[44]

As for the effectiveness of the various Courtauld initiatives, WRAP reports that Courtauld 1 (2005–2009) resulted in the prevention of 1.2 million tonnes of food and packaging waste, with a monetary value of £1.8 billion, and the saving of 3.3 million tonnes of CO_2e.[45] Courtauld 2 (2010–2012) resulted in an estimated reduction of 1.7 million tonnes of waste with a monetary value of £3.1 billion and equates to a reduction of 4.8 million tonnes of CO_2e.[46] Courtauld 3 (2012–2015) was reported to have met its manufacturing and retail target of a reduction in grocery ingredient, product, and packaging waste by 3 percent, equating to 219,000 tonnes of food and packaging waste prevented, representing a CO_2e saving of 555,000 tonnes.[47]

In considering the efficacy of regulatory options for FLW minimization, the question logically arises whether greater results might have been obtained by a mandatory regulatory regime, rather than the voluntary regime which is in place.

One attempt at state regulation of food waste in the UK, the so-called "Food Waste Bill 2015–16" was introduced as a Private Members' bill by an opposition MP, Kerry McCarthy, on 9 September 2015. The bill failed, however, to make any progress during the 2015–16 session of Parliament and fell into abeyance. If enacted, it would have required the Secretary of State for the Environment, Food and Rural Affairs to make provisions for:

> a scheme to establish incentives to implement and encourage the observance of the food waste reduction hierarchy; to encourage individuals, businesses and public bodies to reduce the amount of food they waste; and to require large supermarkets, manufacturers and distributors to reduce their food waste by no less than 30 per cent by 2025, to enter into formal agreements with food redistribution organisations, and to disclose levels of food waste in their supply chain.[48]

The bill contained a number of the features which were included in the French Supermarket Waste Ban Law, discussed in the next section.

The state's recycling minister, Rory Stewart, expressed support for the bill's core principles, but indicated that the "threat of future legislative action" meant that its aims could be achieved through voluntary schemes.[49] The sponsor of the bill indicated that it would remain on the table "as a much-needed regulatory back-up plan if C2025 fails to deliver."[50]

4.4.2 Direct regulation: the EU

The EU does not yet have comprehensive legislation dealing with food waste minimization. With Commission Regulation (EC) No 1221/2008 of 5 December 2008,[51] the European Commission approved the phasing out of previous regulations enforcing certain sizes and shapes of fruit and vegetables sold at retail. This legislative change reduces the emphasis on aesthetic requirements for many fruits and vegetables and so aims to prevent the unnecessary discarding of produce that are aesthetically imperfect but otherwise edible.[52]

In anticipation of EU-wide action, both France and Italy have introduced direct legislation dealing with some aspects of food waste minimization.

4.4.2.1 France

On December 9, 2015, the French National Assembly adopted legislation prohibiting supermarkets from throwing food away or making unsold food unfit for consumption through the addition of chemicals, in an act known as the Supermarket Waste Ban Law.[53] Under the law, supermarkets measuring over 400 square meters must contract one or more other organizations to redistribute unused food from their operations. Noncompliance with these measures can attract fines of up to €75,000.

The legislation had originally passed through the National Assembly in May 2015 as Article 103 of an Energy Bill, but in August 2015 the Constitutional Council ruled that this article was procedurally invalid as it had been added as an amendment during the bill's second reading. Reacting to this court decision, Ségolène Royal, Minister for Ecology and Sustainable Development, called upon the retail sector to adopt the legislative measure on a voluntary, contractual basis, threatening publicly to "out" companies who did not want to take part.[54] However, with bipartisan support, the bill was reintroduced by Guillaume Garot, the former Minister for the Food Industry, on February 11, 2016.[55]

This piece of legislation had been formulated as an element of a larger national food waste policy released in April 2015.[56] The policy had 36 elements, the first of which was to "set into law a hierarchy of preferable actions to fight food waste."[57] The legislation envisaged that by 2025 any organization producing waste above a given threshold will be required to recover edible food according to its highest possible use on the hierarchy, in the following order: "human consumption, animal feed, industrial uses, anaerobic digestion, and composting."[58] Fines were proposed if food ended up in a lower place on the hierarchy than it deserved—for example, if food fit for human consumption was used for animal feed or composted.

The policy hoped to create far-reaching results by involving public education and expecting the widespread promulgation of best practice, as well as envisaging that the French legislation might become part of a Europe-wide food waste code.[59]

4.4.2.2 Italy

Until 2016, Italy had a whole range of laws addressing aspects of food waste minimization. A Legislative Decree of 1992 dealt with labeling, clarifying the difference between expiration and "best-before" dates.[60] Two decrees dealing with the donation of food to charities[61] culminated in the Good Samaritan Law, which entered into force on July 16, 2003. This law, modeled on equivalent legislation in the USA, limited the liability of food companies in relation to products which they donate to charities.[62] The Consolidated Environmental Decree (Testo Unico Ambientale) of 2006 introduced norms concerning waste management and the remediation and treatment of polluted sites.[63] The 2013 Stability Law specified that both donors and beneficiaries of unsold surplus food products must guarantee the proper preservation, transportation, storage, and use of food.[64]

On August 2, 2015, Italy became the second European country to introduce a supermarket waste law. The law made it easier for companies and farmers to donate food to charities and encouraged greater use of "doggy bags" (renamed "family bags") at restaurants. It also allowed stores to donate mislabeled food products if the expiration date and allergy information are properly indicated. The law was a joint production of the Italian National Plan for Food Waste Prevention (Piano nazionale di prevenzione dello spreco alimentare), which the Ministry of Environment began to formulate in 2013, the Zero Waste Charter, launched in 2013 by Last Minute Market, a collector and distributor of surplus food, and the Municipality of Sasso Marconi (Bologna). The aim of these initiatives was to recover products discarded along the entire agro-food supply chain and to redistribute them to people living below the minimum income level, as well as to "change the rules governing public contracts for food and catering services so as to favour firms that guarantee the free redistribution of recovered food."[65]

On September 14, 2016, The Law against Food Waste came into effect.[66] The law's objective is "to reduce waste [at] each of the stages of production, processing, distribution and administration of food, pharmaceuticals and other products" through the implementation of enumerated priorities.[67] These priorities are:

- Promoting the recovery and donation of food surpluses
- Promoting the recovery and donation of pharmaceuticals and other products

- Contributing to limit the negative impacts on environment and natural resources, reducing the production of waste and promoting reuse and recycling to extend products' life cycles
- Contributing to the achievement of the general objectives set by the "National Waste Prevention Program" and the "National Food Waste Prevention Plan," contributing to the "reduction of the amount" of biodegradable waste for landfill sites
- Contributing to information, consumer awareness[68]

The law provides for the donation of food, agricultural, and agri-food goods that remain unsold or discarded from the food supply chain for commercial or aesthetic reasons, or due to proximity to the expiry date, as well as food products that have passed the date of minimum durability within which packaging integrity and suitable storage conditions are guaranteed.[69]

The donated goods are mandated to be provided to the poorest citizens, but if not suitable for human consumption should be used for animal consumption and/or composting.[70]

The law further ensures that receiving associations can collect free agricultural products directly from farmers. These donations are carried out by and under the responsibility of the receiving association or a non-profit organization.

The Italian law on the donation of food waste, unlike the French Law n° 2016-138, does not impose the obligation of food waste donation upon processors and supermarkets, but seeks to establish incentives for donation, as well as to simplify the process of donation.

4.4.2.3 Germany

Germany has no food waste legislation comparable to that in France and Italy.[71] Not only does it lack a "doggy bag" legislation, but it also criminalizes "dumpster diving" (opening commercial garbage containers and collecting food items).[72]

Food waste is addressed in some other, more subtle ways, however, such as in the 2013 Waste Prevention Programme administered by the Federal Government with the participation of the Federal states. The main federal law associated with waste is the 2012 "Kreislaufwirtschaftsgesetz" (Law on Life-Cycle Management), which includes regulations on the prevention, recycling, and disposal of waste and waste management measures. At the federal level, the Pollution Control Act implies food waste minimization,[73] and at the provincial level, the food services in correctional institutions are regulated in Baden-Württemberg and Brandenburg.[74]

The main German food waste minimization initiative that has attracted attention is a proposal by the Minister of Food and Agriculture, announced in March 2016, to replace expiration dates with smart packaging, including the microchipping of dairy products to track freshness.[75]

4.4.2.4 Netherlands

Similarly to Germany and most other EU Members, the Netherlands does not have any direct legislation dealing with food waste minimization; it merely implements the various EU regulations on marketing standards, food contamination, import controls, phytosanitary controls, food hygiene, novel food (GMOs), the provision of food information, fishery quotas, and the use of by-products.[76] However, some researchers claim that the Netherlands applies a very strict implementation of these regulations,[77] which may exacerbate the food wastage problem in that country. In the absence of legislation, a number of organizations voluntarily collect and distribute waste food.[78]

4.4.3 USA

Each of the US states has introduced legislation limiting the liability of food donors in cases of foodborne illnesses carried by their donations.[79] Some states have relaxed the strict liability rules in civil actions, whereas others have also included immunities under criminal law. In 1990, the US Congress attempted to address the issues of liability and a lack of uniformity by developing a federal Model Good Samaritan Food Donation Act.[80] This act did not have the force of law and was only adopted by one state. Endeavoring, in May 1996, to give the Model Act the force of law, Representative Pat Danner, with the co-sponsorship of Representative Bill Emerson, both of Missouri, introduced H.R. 2428, the Model Good Samaritan Food Donation Act. Representative Emerson died on June 22, 1996, before the final passage of the bill, and in his memory, Congress amended the title of the statute to "The Bill Emerson Good Samaritan Food Donation Act." This was signed into law by President Clinton on October 1, 1996,[81] who noted that the complex web of inconsistent state legislation had acted as a hindrance to food donation.[82]

The Bill Emerson Act exempts a donor from liability arising from an injury caused by a food donation made in good faith, excluding liability for acts constituting gross negligence or for intentional misconduct.[83] The Act states that civil or criminal liability shall not arise from the "nature, age, packaging, or condition"[84] of the donated items as long as the donated item is either an "apparently wholesome food"[85] or an "apparently fit grocery product"[86] donated in good faith to a non-profit organization and distributed to needy individuals. The definition of "food" is quite broad because it includes "any raw, cooked, processed, or prepared edible substance, ice, beverage, or ingredient used or intended for use in whole or in part for human consumption."[87]

The Bill Emerson Act exempts gleaners and non-profit organizations from liability. A "gleaner" is "a person that harvests an agricultural crop that has been donated by the owner for either free distribution to the needy directly or to a non-profit organization for ultimate distribution to the needy."[88] A "non-profit organization" can be either an incorporated or unincorporated entity that "operat[es] for religious, charitable, or educational purposes; and does not provide net earnings to, or operate in any other manner that inures to the benefit of, any officer, employee, or shareholder of the entity."[89]

The Federal Food Donation Act 2008[90] supplemented the Bill Emerson Act by encouraging federal agencies to donate excess food to non-profit organizations. Federal contracts for the purchase of food valued at over $25,000 had to make a provision for contractors to donate apparently wholesome excess food to non-profit organizations.

Discussion of food waste legislation has been revived in the USA in recent years. On March 23, 2016, Congresswoman Chellie Pingree (Maine) introduced the Food Recovery Act (HR 4184) and the Food Date Labeling Act (HR 4184) into the House of Representatives. Senator Richard Blumenthal (Connecticut) introduced equivalent legislation into the Senate.[91] None of this legislation, however, has progressed further than the committee stage in Congress. The Food Recovery Acts sought to establish an Office of Food Recovery to oversee the country's efforts to reduce food waste as well as to implement education campaigns in schools and for consumers on food waste. The office would have additionally worked to encourage donations of so-called "ugly" produce, which did not meet high aesthetic standards, to schools. The House bill included tax incentives for donating uneaten food and several other anti-waste measures, but the Senate bill did not include the latter, as they were included in the omnibus spending bill that Congress passed in December 2016.[92]

4.4.4 *Japan*

The Promotion of Utilization of Recyclable Food Waste Act (or Food Recycling Law) came into force in May 2001, driven by food security concerns in an economy that largely imports its food.[93] The Food Recycling Law defines food waste as food materials which are disposed of after being served or without being served or as food and materials which are not able to be served but can be created as a by-product of manufacturing, processing, and cooking processes.[94] The Food Recycling Law mandates the registration of recycling operators and provides a certification system for recycling business plans for food-related businesses. The registration system identifies business operators which seek to conduct their recycling business according to government norms.[95]

Registration secures preferential treatment under the Feed Safety Law and Fertilizer Control Act, eliminating the obligation to notify sales and production of feed and fertilizer, and it enables special treatment under the Waste Disposal and Public Cleansing Law, eliminating the requirement for a work permit for the transportation of municipal solid waste.

An approved business plan for the use of feed and fertilizer from recycled food materials also receives special treatment under the Feed Safety Law, Fertilizer Control Act, and the Waste Disposal and Public Cleansing Law. Under this system, food-related businesses can expect stable supplies of primary products from agriculture, livestock, and fisheries and recycling businesses.[96]

The Food Recycling Law requires revision every 5 years and was amended in 2007, requiring operators with more than 100 tonnes of annual food waste generation to produce status reports annually on the generation and recycling of their food waste. On the basis of these reports, the Ministry of Agriculture, Fisheries and Forestry (MAFF) and the Ministry of Environment (MoE) have established the target value for the control of food waste generation. For a 2-year period, from April 2012, they set target levels for 16 industry groups that dispose of edible parts of food due to overproduction, and since April 2014 have widened the scope to cover 26 industry groups. In October 2014, consumer education and strengthening cooperation amongst local governments were also adopted as future focuses for the legislation.[97]

4.5 Editors' note

This chapter address the question: How can resources from food production and food waste and food loss be saved to achieve more sustainability?

Here, Blakeney provides a comprehensive comparative law analysis of food waste minimization regulation, with several examples that show how various countries and regions around the world address FLW. From the perspective of this book, food waste regulation[98] is a key tool for promoting agroecology and agrobiodiversity: if the law creates the conditions to minimize food waste, it would be possible to implement different food production techniques: less-intensive and aimed more effectively to protect biodiversity (see Figure 4.1). Therefore, the function of food waste regulation is complementary and reciprocal to agroecology and agrobiodiversity regulation.

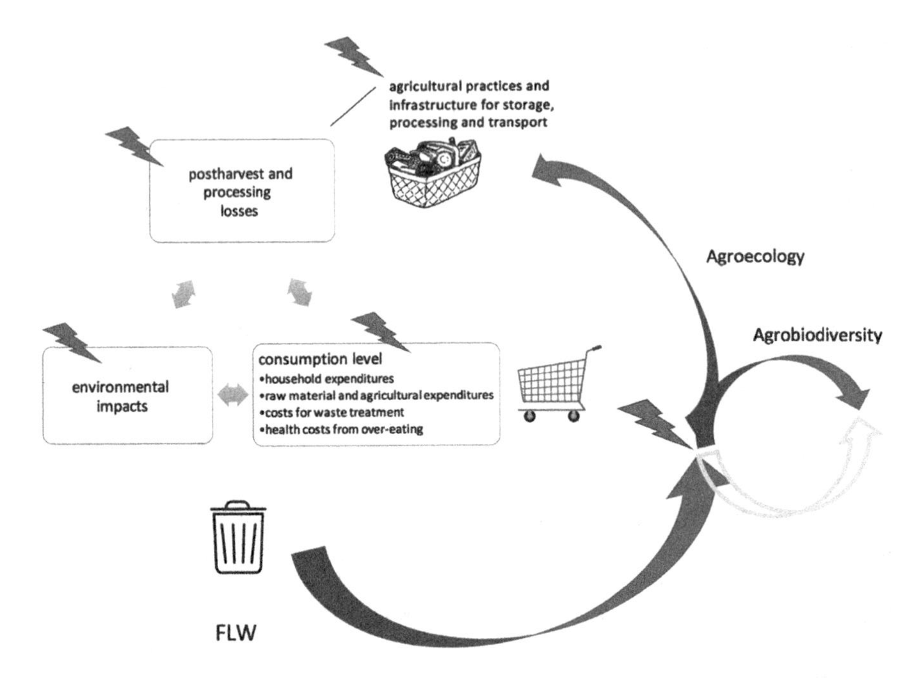

Figure 4.1 Points of regulation to reduce food loss and waste. This figure illustrates the food loss and waste described in Chapter 4 and identifies the regulatory points of attack (represented by lightning bolts). The failure to come to complete cycles in this visual representation reveals the path toward loss, rather than a recycling of resources.

Notes

1. J. Gustavsson, C. Cederberg, U. Sonesson et al., *Global Food Losses and Food Waste: Extent, Causes, and Prevention* (Rome: Food and Agriculture Organization of the United Nations, 2011).
2. United Nations Secretary-General's High-level Panel on Global Sustainability, *Resilient People, Resilient Planet: A Future Worth Choosing* (New York: United Nations, 2012), 12.
3. FAO, *Food Wastage Footprint, Full-Cost Accounting: Final Report* (Rome: Food and Agriculture Organization of the United Nations, 2014); FAO, *Food Wastage Footprint. An Environmental Accounting of Food Loss and Waste* (Rome: Food and Agriculture Organization of the United Nations, 2012).
4. FAO, *Food Wastage Footprint: Impacts on Natural Resources* (Rome: Food and Agriculture Organization of the United Nations, 2013).
5. Maria Rosaria Torrisi, *Food Waste in Australia* (2014), www.futuredirections.org.au/publication/food-waste-in-australia/.

6. United Nations Secretary-General's High-level Panel on Global Sustainability, *Resilient People*, at 36; see also J. Lundqvist, C. de Fraiture, D. Molden. *Saving Water; From Field to Fork: Curbing Losses and Wastage in the Food Chain*, SIWI Policy Brief (Stockholm: Stockholm International Water Institute, 2008), 18.
7. M. Bond, T.Meacham, R. Bhunnoo, T.G. Benton, *Food Waste within Global Food Systems*, A Global Food Security Report (2013), www.foodsecurity.ac.uk; World Bank, Natural Resources Institute, FAO, *Missing Food: The Case Postharvest Grain Losses in Sub-Saharan Africa* (Washington, DC, The World Bank, 2011).
8. Waste and Resources Action Programme, *Understanding Food Waste* (Banbury: Waste and Resources Action Programme, 2007); B. Lipinski, C. Hanson, J. Lomax, L. Kitinoja. R. Waite, T. Searchinger, Reducing food loss and waste, Working Paper (Washington, DC: World Resources Institute, 2013); J. Parfitt, M. Barthel and S. Macnaughton, Food waste within food supply chains: quantification and potential for change to 2050, 365 *Philosophical Transactions of the Royal Society B: Biological Sciences* 3065 (2010).
9. D. Pimentel, Environmental and social implications of waste in US agriculture and food sectors, 3 *Journal of Agricultural Ethics* 5 (1990); E. Morgan, *Fruit and Vegetable Consumption and Waste in Australia* (Carlton South, Australia: Victorian Health Promotion Foundation, 2009); Parfitt et al., Food waste within food supply chains.
10. Regulation (EC) No 178/2002 of the European Parliament and of the Council of 28 January 2002 laying down the general principles and requirements of food law, establishing the European Food Safety Authority and laying down procedures in matters of food safety, OJ L 31, 01.02.2002, pp. 1–24.
11. Within the meaning of Council Directives 65/65/EEC, OJ 22, 9.2.1965, p. 369 and 92/73/EEC, OJ L 297, 13.10.1992, p. 8.
12. Within the meaning of Council Directive 76/768/EEC OJ L 262, 27.9.1976, p. 169. Directive as last amended by Commission Directive 2000/41/EC (OJ L 145, 20.6.2000, p. 25).
13. Within the meaning of Council Directive 89/622/EEC. OJ L 359, 8.12.1989, p. 1. Directive as last amended by Directive 92/41/EEC) (OJ L 158, 11.6.1992, p. 30).
14. European Regulation (EC) No 178/2002.
15. Ibid.
16. FAO, *Regional Strategic Framework Reducing Food Losses and Waste in the Near East and North Africa Region* (Cairo: Food and Agriculture Organization of the United Nations, 2015), 7.
17. Ibid.
18. Andrew Parry, Paul Bleazard, Koki Okawa, Preventing food waste, Case Studies of Japan and the United Kingdom, OECD Food, Agriculture and Fisheries Papers, No. 76, OECD, Paris (2015), 27.
19. FUSIONS, Definitional framework for food waste, full report (July 3, 2014), www.eu-fusions.org/phocadownload/Publications/FUSIONS%20Definitional%20Framework%20for%20Food%20Waste%202014.pdf.
20. Ibid., at 6.
21. FAO, The Consumer Goods Forum (CGF), FUSIONS, United Nations Environment Programme (UNEP), World Business Council for Sustainable Development (WBCSD), WRAP (The Waste and Resources Action Programme).

22. Food loss and waste accounting and reporting standard, FLW Protocol Steering Committee, www.wri.org/sites/default/files/FLW_Standard_final_2016.pdf.
23. Ibid., at 140.
24. Published as J. Gustavsson, C. Cederberg, U. Sonesson et al., *Global Food Losses and Food Waste: Extent, Causes and Prevention* (Rome: Food and Agriculture Organization of the United Nations, 2011), www.fcrn.org.uk/research-library/global-food-losses-and-food-waste-extent-causes-and-prevention.
25. Ibid. See also Ali Chalak, et al., The global economic and regulatory determinants of household food waste generation: a cross-country analysis, 48 *Waste Management* 418 (2016).
26. Ibid., at v.
27. Parfitt, et al., Food waste within food supply chains.
28. N. Gunningham and J. Rees, Industry self-regulation: an institutional perspective, 19 *Law & Policy* 363 at 364 (1997).
29. See G. Teubner, Global Bukowina: legal pluralism in the world society, in G. Teubner (ed.), *Global Law without a State* (Aldershot, UK: Dartmouth, 1997), 3–28; J. Braithwaite and P. Drahos, *Global Business Regulation* (Cambridge, UK: Cambridge University Press, 2000); N. Gunningham and D. Sinclair, Regulatory pluralism: designing policy mixes for environmental protection, 21 *Law & Policy* 49 (1999); G. Teubner, Global private regimes: neo-spontaneous law and dual constitution of autonomous sectors in world society?', in K.H Ladeur (ed.), *Public Governance in the Age of Globalization* (Farnham, UK: Ashgate, 2004), 71–87; N. Gunningham, Corporate environmental responsibility: law and limits of voluntarism, in D. McBarnet, A. Voiculescu, T. Campbell (eds.), *The New Corporate Accountability: Corporate Social Responsibility and the Law* (Cambridge, UK: Cambridge University Press, 2009), 476–500.
30. See Australia, Taskforce on Industry Self-regulation, Report on industry self-regulation in consumer markets (2000), https://archive.treasury.gov.au/documents/1131/HTML/docshell.asp?URL=01_prelims.asp.
31. See M. Blakeney and S. Barnes, Industry self-regulation an alternative to deregulation? Advertising a case study, 5 *UNSWLJ* 132 (1982); V. Haufler, *A Public Role for the Private Sector: Industry Self-regulation in a Global Economy* (Washington, DC, Carnegie Endowment for International Peace, 2001).
32. See N. Gunningham, R.A. Kagan, D. Thornton, *Shades of Green: Business, Regulation, and Environment* (Stanford, CA: Stanford University Press, 2003).
33. J. Howard, J. Nash, J. Ehrenfeld, Industry codes as agents of change: responsible care adoption by chemical companies, 8 *Business Strategy and the Environment* 281 (1999); H.S. Dashwood, Sustainable development and industry self-regulation: developments in the global mining sector, 53 *Business & Society* 551 (2014); S. Nysten-Haarala, E. Klyuchnikova, H. Helenius, Law and self-regulation: substitutes or complements in gaining social acceptance? 45 *Resources Policy* 52 (2015).
34. B. Cashore, G. Auld, S. Bernstein, C. McDermott, Can non-state governance "ratchet up" global environmental standards? Lessons from the forest sector, 16 *Review of European Community and International Environmental Law* 158 (2007).
35. Institute of Medicine, *Food Marketing to Children and Youth: Threat or Opportunity?* (Washington, DC, National Academies Press, 2006); C. Hawkes, Regulating food marketing to young people worldwide: trends and policy drivers, 97 *American Journal of Public Health* 1962 (2007).

36. C. Hawkes, Self-regulation of food advertising: what it can, could and cannot do to discourage unhealthy eating habits among children, 30 *Nutrition Bulletin* 374 (2005); L.L. Sharma, S.P. Teret, K.D. Brownell, The food industry and self-regulation: standards to promote success and to avoid public health failures, 100 *American Journal of Public Health* 240 (2010).
37. M.M. Mello, J. Pomeranz, P. Moran, The interplay of public health law and industry self-regulation: the case of sugar-sweetened beverage sales in schools, 98 *American Journal of Public Health* 595 (2008).
38. L.G. Smithers, J.W. Lynch, T. Merlin, Industry self-regulation and TV advertising of foods to Australian children, 50 *Journal of Paediatrics and Child Health* 386 (2014).
39. T.F. Babor, Z. Xuan, D. Damon, A new method for evaluating compliance with industry self-regulation codes governing the content of alcohol advertising, 37 *Alcoholism: Clinical and Experimental Research* 1787 (2013).
40. The Courtauld Commitment 2025, Cutting the cost of food and drink, www.wrap.org.uk/content/what-courtauld-2025.
41. The Courtauld Commitment 2025 to transform UK food and drink (March 15, 2016), www.wrap.org.uk/content/courtauld-commitment-2025-transform-uk-food-and-drink.
42. Retailers: Aldi, Asda, Central England Co-operative, Lidl, Marks & Spencer, Morrisons, Musgraves, Sainsbury's, Tesco, Co-operative Food, and Waitrose; brands and manufacturers: ABF UK Grocery Group, ARLA, Birds Eye UK, Coca-Cola Enterprises, Heineken, Nestlé UK and Ireland, Premier Foods, Unilever, and Warburtons; hospitality and food service: apetito, Bidvest, Compass, Greene King Retail, KFC, OCS, Pizza Hut, Sodexo UK & Ireland; local authorities: 23 authorities including the London Waste and Recycling Board, representing more than 42% of the UK's population; trade and sector organizations, government, and academia: British Hospitality Association, British Retail Consortium, Chilled Food Association, Dairy UK, Food & Drink Federation, Food Standards Agency, Institute of Hospitality, Sustainable Restaurant Association, and WWF. Ibid.
43. The Courtauld Commitment 2025, 25 *Food Packaging Bulletin* 11 at 12 (2016).
44. The Courtauld Commitment 2025 to transform UK food and drink.
45. The Courtauld Commitment 2025, Cutting the cost of food and drink.
46. Ibid.
47. Courtauld Commitment 3 delivers over £100 million business savings by reducing food waste over three year period, www.wrap.org.uk/content/courtauld-commitment-3-delivers-over-%C2%A3100-million-business-savings-reducing-food-waste-over-.
48. Food Waste (Reduction) Bill 2015–2016, https://services.parliament.uk/bills/2015-16/foodwastereduction.html.
49. The Food Foundation, Food Waste Reduction Bill, www.foodfoundation.org.uk/food-waste-reduction-bill/.
50. CIWM, Food Waste Reduction Bill: what's next?, https://ciwm-journal.co.uk/food-waste-reduction-bill-whats-next/.
51. Commission Regulation (EC) No 1221/2008 of 5 December 2008 amending Regulation (EC) No 1580/2007 laying down implementing rules of Council Regulations (EC) No 2200/96, (EC) No 2201/96 and (EC) No 1182/2007 in the fruit and vegetable sector as regards marketing standards, OJ JOL_2008_336_R_0001_01.

52. The current list of fruit and vegetables impacted are: apricots, artichokes, asparagus, aubergines, avocadoes, beans, Brussels sprouts, carrots, cauliflowers, cherries, courgettes, cucumbers, cultivated mushrooms, garlic, hazelnuts in shell, headed cabbage, leeks, melons, onions, peas, plums, ribbed celery, spinach, walnuts in shell, water melons, and witloof chicory. The exception from marketing standards could be extended to another ten products such as apples, citrus fruit, kiwifruit, lettuces, peaches and nectarines, pears, strawberries, sweet peppers, table grapes, and tomatoes. See European Commission, food waste, promote good practices, https://ec.europa.eu/food/safety/food_waste/good_practices/policy_awards_certification_en.
53. Proposition de loi relative à la lutte contre le gaspillage alimentaire, December 9, 2015.
54. Niamh Michail, France's food waste law scrapped on a technicality, August 28, 2015, www.foodnavigator.com/Article/2015/08/28/France-s-food-waste-law-scrapped-on-a-technicality.
55. Loi n° 2016-138 du 11 février 2016 relative à la lutte contre le gaspillage alimentaire, www.legifrance.gouv.fr/affichTexte.do?cidTexte=JORFTEXT000032036289&categorieLien=id.
56. *Lutte contre le gaspillage alimentaire: propositions pour une politique publique*, Mission parlementaire menée par Guillaume Garot sur la lutte contre le gaspillage alimentaire, April 14, 2015, https://agriculture.gouv.fr/telecharger/72171?token=17ca3443c44991fa1f25c901dc7a66ce.
57. See Marie Mourad, *France moves toward a national policy against food waste* (Paris: Natural Resources Defense Council, 2015), 4, www.nrdc.org/sites/default/files/france-food-waste-policy-report.pdf.
58. Ibid., at 5.
59. Ibid., at 11.
60. Legislative Decree No. 109/1992 implementing EU Directives nos. 89/395/CEE and 89/396/CEE concerning the labelling, packaging, and advertising of foodstuffs.
61. Legislative Decree, 4 December 1997, No. 460/1997; Decree of the President of the Italian Republic, 26th of October 1972, n. 633.
62. Law no. 155/2003.
63. Consolidated Environmental Decree No. 152/2006 (amended in 2010).
64. Law No. 147 of 27 December 2013.
65. Zero Waste Charter, www.unannocontrolospreco.org/images/CartaSprecoZero2.1.pdf.
66. Law No. 166/2016, concerning provisions on the donation and distribution of food and pharmaceutical to limit food waste.
67. Food Law Latest, The Italian Law against food waste, https://foodlawlatest.com/2016/11/26/the-italian-law-against-food-waste/.
68. Ibid.
69. Ibid.
70. Ibid.
71. FUSIONS, Review of current EU Member States legislation and policies addressing food waste, www.eu-fusions.org/phocadownload/Reports/GERMANY%20FULL%20pdf.pdf.

72. M. Rombach and V. Bitsch, Food movements in Germany: slow food, food sharing, and dumpster diving. 18 *International Food and Agribusiness Management Review* (2015), www.ifama.org/resources/Documents/v18i3/Rombach-Bitsch.pdf.
73. Gesetz zum Schutz vor schädlichen Umwelteinwirkungen durch Luftverunreinigungen, Geräusche, Erschütterungen und ähnliche Vorgänge (Bundes-Immissionsschutzgesetz: BImSchG) (Act on the prevention of harmful effects on the environment caused by air pollution, noise, vibration, and similar phenomena) (§ 5 Abs. 1 Nr. 3).
74. Verpflegungsordnung für die Justizvollzugsanstalten des Landes Baden-Württemberg; Verpflegungsordnung für die Justizvollzugsanstalten des Landes Brandenburg, Rundverfügung der Ministerin der Justiz.
75. Germany plans "smart" packaging to cut food waste, www.euractiv.com/section/agriculture-food/news/germany-plans-smart-packaging-to-cut-food-waste/.
76. FUSIONS, the Netherlands: country report on national food waste policy, www.eu-fusions.org/phocadownload/country-report/NETHERLANDS%2023.02.16.pdf, table 2.
77. Y. Waarts, M. Eppink, E.B. Oosterkamp et al., Reducing food waste: obstacles experienced in legislation and regulations, LEI report 2011-059 (Wageningen, the Netherlands: Wageningen University and Research, 2011).
78. S. van der Meulen, G. Boin, Food waste and donation policies in France and the Netherlands: initiatives to reduce food waste, 4/15 *eFOOD-Lab international* 27 (2015).
79. See James Haley, The legal guide to the Bill Emerson Good Samaritan Food Donation Act, *Arkansas Law Notes* 1448 (2013).
80. Public Law No. 101-610 (November 16, 1990), §§ 401–02, 104 Stat. 3127, 3183–85 (1990).
81. Public Law No. 104-210 (October 1, 1996), 110 Stat. 3011, 3011 (codified at 42 USC § 1791 (2011)).
82. Presidential Signings, Statement on Signing HR 2428, 32 *Weekly Compilation of Presidential Documents* 1943 (October 7, 1996).
83. Public Law No. 104-210, 42 USC § 1791.
84. Ibid., at § 1791(c)(1).
85. Ibid., at § 1791(b)(2): "means food that meets all quality and labeling standards imposed by Federal, State, and local laws and regulations even though the food may not be readily marketable due to appearance, age, freshness, grade, size, surplus, or other conditions."
86. Ibid., at § 1791(b)(1): "means a grocery product that meets all quality and labeling standards imposed by Federal, State, and local laws and regulations even though the product may not be readily marketable due to appearance, age, freshness, grade, size, surplus, or other conditions."
87. Ibid., at § 1791(b)(2).
88. Ibid., at § 1791(b)(5).
89. Ibid. at § 1791(b)(9).
90. S.2420, Federal Food Donation Act of 2008, Public Law No. 110-247 (06/20/2008).
91. S.3108, Food Recovery Act of 2016, Introduced in Senate (06/29/2016); S.2947, Food Date Labeling Act of 2016, Introduced in Senate (05/18/2016).

92. H.R. 2029, Consolidated Appropriations Act 2016, 12/18/2015, Became Public Law No: 114-113.
93. F. Marra, Fighting food loss and food waste in Japan, MA thesis, Leiden University, 2013.
94. A. Parry, P. Bleazard, K. Okawa, Preventing food waste: case studies of Japan and the United Kingdom, OECD Food, Agriculture, and Fisheries Papers, No. 76, OECD Publishing, Paris (2015), 9–10.
95. Ibid., at 10.
96. Ibid.
97. Ibid., at 25.
98. On the general profiles of food waste regulation, see S.J. Morath, Regulating food waste, 48 *Texas Environmental Law Journal* 239 (2018); for a European comparative approach, see L.G. Vaque, French and Italian food waste legislation, 12 *European Food and Feed Law Review* 224 (2017).

chapter five

Indigenous peoples and agrobiodiversity in Africa

Rosemary E. Agbor and Wele Elangwe

Contents

5.1 Introduction

In the absence of the technologically focused systems of the western world, indigenous practices have been the backbone of the preservation of plant and animal life across the African continent from time immemorial. Such indigenous biodiversity-preserving practices have steadily passed down from one generation to the next like folklore. Today, some of these practices face the risk of slow extinction with the increasing adoption of modern, scientific research-based practices; while others, tried, tested, and proven effective even in today's world, continue to thrive.

The inhabitants of the western and central parts of the African continent share various cultural and physical similarities that don't conform to the boundaries of the countries they live under, so it's no surprise that some indigenous conservation practices in this region transcend national or provincial boundaries. While some of these practices have received international praise for their resource-conservation capabilities, some others have earned much criticism, and yet others are still tentatively moving toward the path of success. However, whether they are good, bad, or somewhere in between, indigenous efforts and activities affecting plant, soil, and animal life in Africa merit discussion in any educated discourse on biodiversity. This chapter will discuss three case studies from distinct indigenous groups of the African continent:

- *tassa/zai* farmers of the Sahel region
- the Pygmies of the Congo Basin
- the Mboboro pastoralists of the North-West Region of Cameroon.

This chapter provides a regional perspective of agrobiodiversity, with case studies of three indigenous groups from Sahel, Congo Basin, and Cameroon. The impacts of indigenous peoples on biodiversity conservation efforts in Africa, though often overlooked, cannot be overemphasized. Traditional or indigenous knowledge-based agricultural practices may have positive or negative effects on agrobiodiversity. One indigenous practice with a positive impact is a traditional irrigation technique known as *zai* or *tassa*, practiced in the Sahel region, the success of which has shown that community-based food security need not depend upon expensive and unsustainable imported western technology. On the other side of the sustainability spectrum, however, lies the indigenous practice of hunting and eating "bushmeat," predominantly practiced by the indigenous Pygmy tribes of Central Africa, which has negatively affected the conservation efforts of many protected species as well as fostered the spread of certain diseases such as Ebola. However, despite the negative effects some indigenous practices may have, when communities come together to find solutions, positive results can still emerge. One such case lies with the Mbororo people, an indigenous group in Cameroon comprised predominantly of cattle herders. A scarcity of grazing land has fostered conflict between the Mbororos and local farmers using the land for agriculture. But, through dialogue, a mutually beneficial solution has been proffered between these two groups in the form of alliance farming. Interestingly, another good has emerged from this resolution as cattle manure is now also being used to produce biogas for fuel—further enhancing agrobiodiversity conservation efforts.

5.2 *Case study: Conservation to combat desertification in the Sahel: the* zai/tassa *system of water management*

5.2.1 *Overview*

The Sahel refers to a transitional zone covering an area of about 3,050,000 square kilometers (1,180,000 square miles), covered by semi-arid grasslands, savannas, steppes, and thorny shrublands, lying between the Sahara Desert and the wetter lands of Southern Africa.[1] It includes parts of Senegal, Mali, Burkina Faso, Algeria, Niger, Mauritania, Nigeria, Chad, Cameroon, Central African Republic, Sudan, South Sudan, Eritrea, and Ethiopia.[2] In recent years, the Sahara has been encroaching on the formerly arable lands of the Sahel; this desertification has been attributed to decreased rainfall and unsustainable agricultural practices in the region.[3]

For the inhabitants of the Sahel, water scarcity and land degradation due to various factors, including larger climatic variations as well as local human activities, are harsh realities, making food production and natural resource sustainability a larger challenge than in many other regions of the world. To mitigate the adverse effects of desertification and water scarcity, farmers in this region have resorted to unique water- harvesting techniques, including planting pits called *tassa* (or *zai*, in some countries) to reclaim land which is already or about to be lost to degradation.

Tassa is an indigenous irrigation technique that has been used historically and is currently being used by farmers in the Sahel region of Africa to conserve water and soil. The practice entails planting seeds in small hand-dug circular pits on soil that is not very permeable and filling them with manure and biodegradable waste. The manure attracts termites to digest the organic matter of the soil, making nutrients more easily available to the crops planted or sown in the pits. The holes measure about 20 cm in width and depth and are dug during the dry season. The pits also collect water during the rainy season and store it for the crops to use during the often-harsh dry season. The pits concentrate water and nutrients in one location to increase soil fertility and crop yield.

Although the planting requires heavy labor to dig when the soil is dry, the returns are worth the trouble during the prolonged periods of drought when the crops may otherwise have died because of lack of water. Once planted, the pits can be used again and again, season after season.

5.2.2 *Origins of* tassa/zai

Water-harvesting techniques are commonly used by farmers in the Sahel region, especially in Niger, Burkina Faso, and Nigeria. The modern and improved version of the water-harvesting practice of pit planting can be traced to the Yatenga province of Burkina Faso, following two major failed attempts by international donors to irrigate the land and curb soil erosion in the 1960s and 1970s. Those projects were fashioned solely on foreign techniques and brought more harm than good, leaving the land in a worsened state of degradation.[4] Left without other alternatives, local farmers resorted to the indigenous water-harvesting practices of pit planting to combat desertification and soil erosion.

Along with much of the rest of the African continent, Burkina Faso suffered one of the worst environmental crises in its history in the 1970s and 1980s, when annual rainfall dropped by almost 80 percent. The frequent recurrence of droughts led to difficulties in soil cultivation and harvest failures, spurring mass migrations of populations from the Sahel to the Ivory Coast and other areas of Burkina Faso with better food supply and higher rainfall.[5] Those who remained, however, were faced with the task of identifying novel practices for survival in the face of impending desertification.

Central to those efforts were Yacouba Sawadogo and Ousseni Zorome, two local farmers with no access to modern techniques and tools, who tapped into traditional water conservation practices to reclaim the soil. The father of modern-day *zai*, Sawadogo has been dubbed "the man who stopped a desert using ancient farming" for his innovative technique to rehabilitate severely degraded farmland previously impenetrable by water.[6]

Yacouba Sawadogo, the man who reinvented the indigenous technique of *zai* pits (https://www.lifegate.com/people/news/yacouba-sawadogo-the-man-who-stopped-the-desert)

In 1979, Sawadogo began planting *zai* pits with the sole aim of achieving food sufficiency for his family. He took the traditional practice of digging the *zai* pits and adjusted it by making them a little larger and filling them with manure and other biodegradable waste. Although initially ridiculed, Sawadogo's innovation soon earned great respect when the positive effects became all too evident. The pits brought in unprecedented crop yields, and Sawadogo was also pleasantly surprised to see various tree species sprouting in the areas where the *zai* pits were introduced.

Initially, Sawadogo limited his *zai* pits to the growth of sorghum and millet, but eventually he progressed to growing trees and small shrubs. Within two decades of reinventing the *zai* technique, he had succeeded in raising about 20 hectares of forest in what was once semi-arid desert land. Sawadogo's innovation would eventually gain international recognition for its effectiveness in reversing desertification in the region, earning him an alternative Nobel prize in 2018.[7]

Following the success of his reinvention, Sawadogo went on to share his knowledge with other local farmers. As early as 1984, he began organizing biannual market days where he educated other local farmers on the use of *zai* pits. By the 2000s, his *zai* market days had gained great popularity among farmers near and far; an average *zai* market day attracted farmers from about a hundred villages. Sawadogo's disciple Ousseni Zorome went a step further in 1992 by opening, on a borrowed gravelly piece of farmland, the first *zai* school to train farmers in the practice.[8] His roadside school eventually garnered the interest of the Ministry of Agriculture when his crop-growing efforts on this degraded piece of land proved quite successful. Less than a decade after he started, he had already recorded huge successes: his network included over 20 schools and 1,000 members who continued successfully to rehabilitate borrowed degraded land for food production.[9]

His efforts were copied by a relatively rich businessman Ali Ouedraogo, who traveled around the Gourcy region of Burkina Faso visiting individual farmers and training them on *zai* practices. His students in turn experimented with their own techniques and passed the knowledge on to other local farmers. In these ways, *zai* spread across the region in leaps and bounds.

Zai pits in Burkina Faso (https://www.echocommunity.org/resources/d676d269-5f1f-47f1-812a-ed6d3e253989)

5.2.3 The impact of water-harvesting practices

Zai/tassa has received accolades for its ability to meet the criteria of three types of conservation practices—soil conservation, water conservation, and erosion protection.[10] Inexpensive, yet highly effective, this simple indigenous technique has resulted in major increases in crop yields and helped farmers build resilience against drought and soil degradation. Today, *tassa* is seen as "an integral part of the local farming scene" and "is still spreading at a rate of about two to three hectares per year," according to Ezeanya-Esiobu.[11] The advantages of water harvesting are so numerous that farmers in that region call it "the invention of the century."[12]

5.2.4 Impacts on food scarcity

In an experiment that was conducted, two similar plots of land were used ... one plot of land did not have the *Tassa* technique on it, the other one had *Tassa* technique constructed on it. Then similar grains of millet also were planted on both plots. During harvest time, the plot of land without *Tassa* technique yielded 11 kilograms (24 pounds) of millet per hectare. The plot of land with *Tassa* technique yielded 553 kilograms (1,219 pounds) of millet per hectare.[13]

5.2.5 How Africa can use its traditional knowledge to make progress

Tassa/zai has been successful in rehabilitating degraded land in areas with very limited resources and boosting food production in a region previously plagued by acute food scarcity. Using this practice, farmers have not only been able to improve soil quality and expand their farmlands, but also exponentially to increase their yields per hectare within a short period of time. In Niger's Illela district, for example, cereal yields on fields treated by *zai* rose by an average of 388 kilograms per hectare over a 6-year period.[14] Even when another severe drought hit Niger between 2005 and 2006, villages whose land had been rehabilitated by water-planting techniques were hit less hard than those areas where the techniques had not been adopted.

In Mali, *zai* pits have been attributed to an increase in sorghum yields by an average of 719 kilograms per hectare over a 2-year period.[15] Likewise, in certain areas of Nigeria, *tassa* has been attributed to large increases in food production, effectively raising yields from virtually nothing to "300 to 400 kilograms per hectare in a year of low rainfall, and up to 1,500 kilograms or more per hectare in a good year."[16] In Burkina Faso, *zai* pits have been attributed to a 325 kilograms per hectare increase in yields for millet and sorghum, which remain the dominant crops produced in that region. With the land rehabilitated by the *zai* pits, however, Burkinabe farmers have also been able to expand the production of other crops such as sesame and cotton.

5.2.6 Impacts on biodiversity

Planting pits has changed the topography of the region from dry and barren to lush, rich vegetation. Since the land rehabilitation started in Burkina Faso, there has been a surge in the number of trees grown per hectare. Furthermore, areas treated with the pits produce greater diversity of plant species compared with untreated areas, and trees grown on treated lands are larger than those grown on untreated plots. For example, when Ousséni Zoromé began using *zai* pits in the early 1980s, he reported counting only 9 trees on the 11 hectares of degraded land he started off with. Today, that same piece of land boasts of at least 2,000 trees representing 17 species of plant life, including some species previously on the edge of extinction.[17] For example, since the reintroduction of *zai* pits in Burkina Faso, baobab trees have been preserved from local extinction via regeneration by the *zai* farmers.[18]

Even though the planting pits are dug on dry, barren land, otherwise incapable of any plant growth, research has reported an increase in soil fertility in the areas treated to *tassa/zai* when compared with similar non-treated pieces of land. After the introduction of the planting pits, the soil structure, organic matter, and nitrogen content on the land significantly increased, while the clay content significantly decreased.[19] These changes can be attributed to the manure deposited into the pits to boost soil fertility. *Zai* farmers furthermore practice site-specific crop management by gauging the quality of the land every other year and adding compost as needed.

Concurrent with the growth in plant life has been a resurgence in livestock husbandry in the area, since the former thrives in part on the latter. Before the adoption of *zai*, many farmers in the Niger/Burkina Faso outsourced livestock rearing to Fulani herdsmen. However, with the increased need for manure to feed *zai* pits, many local farmers have now integrated livestock rearing into their cropping activities. The proliferation of trees as a result of *zai* has increased the supply of fodder, making it possible and profitable for farmers to keep their livestock closer to home.

5.2.7 Socio-economic impacts

Pit-planting techniques have been attributed to the alleviation of poverty in the areas in which they have been applied. In the Zodoma Province of Burkina Faso, for example, the number of poor families decreased by 50 percent between 1980 and 2001 following the rehabilitation of about 600 hectares of previously unproductive land.[20] Prior to the introduction of the *zai* pits, severe drought conditions forced the mass migration of families from local villages to neighboring areas in Ivory Coast in search of food, water, and fertile lands.

Since the introduction of *zai*, however, the drought conditions and associated food scarcity and ensuing outward migration, dry wells and barren lands are a thing of the past. In fact, according to the villagers of Ranawa (Zodoma Province), the introduction of *zai* pits in the locality has brought to an end the trend of migration which saw the departure of 49 families (25 percent of the population) between 1970 and 1980 for settlement in neighboring Ivory Coast. According to them, since 1985, not a single family has migrated out of the province due to difficult living conditions.[21] These newly improved conditions have also impacted the flow of human capital: the pattern of young men migrating to urban areas after the harvest season in search of employment is gradually fading as agriculture in the rural areas now provides opportunities for year-round employment.

The vegetation has also provided new economic opportunities for the local farmers. For example, women in the Zinder region of Niger who grow baobab trees are now making substantial earnings (about $210 annually) from the sale of tree leaves which are used for the preparation of a local porridge.[22] Other farmers sell the wood to meet local construction demands, pulling in revenues of about 20,000–30,000 fcfa ($35–$50) per annum.

5.2.8 Conclusion

The success of *zai/tassa* pits has shown that the solutions to some of the problems which plague local communities may lie in their own traditional knowledge base and practices; and stakeholders ought to explore these practices before seeking expensive foreign solutions.[23] In the 1970s and 1980s, the notion of regreening the Sahel region would have been a far-fetched dream. Yet, thanks to the successful reinvention of indigenous farming practices, not only is the region now thriving, but it has also developed into a stable home to some species that were threatened with extinction just a few decades ago.

It bears mentioning that the *tassa/zai* technique has not miraculously solved all the environmental problems that plague the Sahel region, but its contribution to the reconstruction of this region cannot be overemphasized. That notwithstanding, the threat of desertification of the Sahel region persists, and the World Bank and other donor agencies continue to advocate the use of mineral fertilizers to rehabilitate the degraded lands. However, because these mineral fertilizers typically prove too expensive for local farmers to acquire, they are frequently passed over in favor of local, relatively inexpensive interventions. Thus, water-planting practices thrive as methods for the rehabilitation of degraded, barren lands and for preserving the biodiversity of plant and animal life in the Sahel region.

5.3 *Case study: Pygmies and bushmeat hunting and consumption in the Congo Basin*

5.3.1 *Overview*

Pygmy peoples, also referred to as forest peoples, are the largest group of indigenous hunter–gatherers in sub-Saharan Africa, and they predominantly inhabit the main forest blocks in the Congo Basin.[24] Pygmy culture dates back more than 20,000 years, and bushmeat consumption is an integral part of the Pygmy livelihood, as they depend on the forest for their physical as well as cultural survival.[25] They also rely heavily on bushmeat for nutritional, economic, and cultural purposes.[26] Pygmies use bushmeat hunting to supplement their primary source of income, which is often agriculture. In the Congo Basin, an estimated 282 grams of bushmeat are consumed per person per day, and over 3 million tons are harvested in Central Africa each year.[27]

The Baka Pygmies of Cameroon (http://blogs.discovermagazine.com/gnxp/2011/05/the-Pygmies-are-short-because-nature-made-them-so/#.WvsEOIgvzIU)

5.3.2 *Biodiversity in the Congo Basin*

The Congo Basin, located in Central Africa, consists of rich rainforests, rivers, and savannahs. It is endowed with a medley of endangered wildlife such as elephants, chimpanzees, lowland and mountain gorillas, antelopes, bonobos, and buffalo.[28] Furthermore, the region contains approximately 10,000 species of tropical plants, 30 percent of which are unique to the region,[29] and about 1,000 species of birds, 400 other species of mammals, and 700 species of fish that add to its uniquely diverse biome.[30] The basin lies under the territory of six sovereign states: Cameroon, the Central African Republic (CAR), the Democratic Republic of the Congo (DRC), Equatorial Guinea, Gabon, and the Republic of Congo (Congo).[31] Pygmies are present in all these countries, constituting, for example, about 0.4 percent of the total population of Cameroon. Pygmy communities in Cameroon live along its forested borders with Gabon, the Congo and the CAR,[32] deriving their entire livelihoods from the forests.[33] Some historians believe that all of Gabon may even have been originally inhabited by Pygmies.[34]

5.3.3 *Bushmeat consumption in the Congo Basin*

The term "bushmeat" refers to meat procured from wild animals hunted from forests, which aside from being used for personal consumption, is often sold commercially.[35] The bushmeat trade has long been recognized as a severe threat to wildlife populations in the forests of Central Africa and is considered a conservation crisis in the region. In the Congo Basin, urban bushmeat consumption is significant and has become a major conservation problem.[36] Urban populations in Gabon, the DRC, and CAR reportedly consume, on average, 4.7 kilograms per person of bushmeat per year. Given the very significant bushmeat consumption in Central Africa and the either nonexistent (e.g. Gabon, DRC, Congo) or extremely limited (Cameroon, CAR) domestic livestock sector, bushmeat remains a crucial component of food security in the region.[37]

Bushmeat hunting and consumption historically is an indigenous practice of Pygmy people. However, population growth and urbanization, coupled with the presence of logging trucks and improved road infrastructure, have transformed the Pygmies' traditional practice into an all-out commercial endeavor, staged not for subsistence but to feed increased demand and growing regional and urban markets. This increased demand for bushmeat is a threat to Pygmy communities since overhunting may lead to the eradication of the very wildlife that sustains them. With the wildlife gone, their own existence as a forest-dwelling people would disappear as well. In fact, this process is already taking place as forests are ripped apart amid soaring population growth and increased logging and road infrastructure, forcing multitudes of Pygmies into settled lives in big cities and towns so far removed from their historical dwellings in the hearts of forests.

Bushmeat was originally exploited as a low-cost protein alternative to farm-raised meat but has now gained popularity and increased in value, which in turn has attracted indiscriminate hunting, to the point where Central Africa has lost significant amounts of wildlife in recent decades. There has been a general shift from hunting for domestic consumption towards hunting for trade and profit, along with increased access to previously remote areas for widespread hunting. Likewise, the proliferation of guns and other more sophisticated hunting technology has altered traditional hunting behavior and has allowed hunters successfully to hunt more prey.[38] Bushmeat consumption is widespread throughout the Congo Basin and common in both rural and urban areas. In remote and impoverished rural areas where the Pygmies reside, bushmeat is often an essential protein source crucial for local food security, particularly in areas where other protein sources such as livestock and fish are either inaccessible or unaffordable.[39] In these rural areas, cultural factors and social needs also shape bushmeat-hunting activities. Studies have reported that the increased demand for bushmeat can also be attributed in part to its use in cultural and medicinal practices.[40]

Bushmeat advertised for sale.[41]

5.3.4 *International bushmeat trade*

Due to the minimal costs of obtaining bushmeat and the high prices that they can fetch on the market, the high potential for profit by commercial hunting operations has pushed bushmeat into new national and international markets.[42] For example, an estimated five tons of bushmeat are smuggled from Africa to Europe every week.[43] A 2010 study conducted by researchers from the Zoological Society of London estimated that about 270 tons of bushmeat are smuggled annually through a single airport in Paris, intended both for personal consumption by households abroad as well as to supply the lucrative commercial trade, nationally and internationally.[44]

Meanwhile, in the USA, several restrictions have been placed on the importation of bushmeat.[45] In fact, it is illegal to ship, mail, or bring bushmeat into the USA. Moreover, immigration officials are empowered to confiscate and destroy any bushmeat found at the port of entry. Perpetrators may also be fined for attempting to smuggle bushmeat into the country. In spite of these efforts, thousands of pounds of bushmeat continue to be smuggled into the USA every year.[46] In fact, worldwide, wildlife is second only to narcotics among black market trades.[47]

5.3.5 *Effects of bushmeat consumption*

5.3.5.1 *Wildlife depletion*

For both Pygmies and non-Pygmies in the region, bushmeat, along with fish, is a traditional food staple and a significant protein source. However, since hunters typically prefer larger animals, an ecological crisis has emerged. Once the large animals become scarce in an area, hunters begin to deplete the populations of the smaller animals as well.[48]

But the responsibility for these crises is not shared equally among the different types of hunters. Pygmies tend to hunt larger and fewer prey species destined for consumption within their own communities, while non-Pygmies hunt more to sell for profit.[49] Furthermore, non-Pygmies hunt a wider range of species, and on average twice as many animals are taken per square kilometer. In fact, research has shown that the non-Pygmy population may be responsible for 27 times more animals harvested than the Pygmy population in Central Africa.[50] This figure indicates that Pygmy subsistence hunting is less detrimental to the local ecology than the widespread commercial hunting perpetrated by non-Pygmies.

5.3.5.2 *Disease risk*

Bushmeat hunting threatens biodiversity and increases the risk of zoonotic disease transmission (from animals to people).[51] Many countries worldwide, such as the USA and the United Kingdom, have adopted strict rules regarding the importation of bushmeat following disease outbreaks that were linked to it.

The Ebola virus disease (EVD) epidemic that emerged in March 2014 in West Africa was the largest ever recorded, resulting in over 28,600 cases and 11,300 deaths in Guinea, Liberia, and Sierra Leone.[52] EVD is a deadly zoonotic disease transmitted to humans through contact with the blood and other bodily fluids of infected wildlife, such as fruit bats, forest antelopes, and non-human primates.[53] Although several other factors allegedly contributed to the spread of the Ebola virus, bushmeat hunting and consumption was suspected as a potential source of the initial transmission and as a cause of the rapid infection rates of the virus during the crisis.[54]

5.3.5.3 *Ineffective legal systems*

The hunting of bushmeat is regulated in most African countries through wildlife legislation. The majority of bushmeat hunters, however, ignore such restrictions.[55] In many African countries, law enforcement agencies are ill-equipped to administer the regulations, and, in some cases, political instability, poor governance, and corruption may protect offenders from the consequences mandated by the law.[56]

In addition, because the legal system of most of Central Africa states rests principally on Francophone, colonial traditions, all forests, water, minerals—and, in fact, all natural resources—are the legal domain of the national governments.[57] As a result of the weak state management of these resources, local populations are often stripped of benefits. Meanwhile, wealthier, stronger, or elite groups, including private companies who are able to bribe their way through or maneuver the weak governance structures, exploit the natural resources at local levels.[58]

One of the major issues limiting bushmeat trade regulation is land ownership. This is especially with regard to the difference in modern legal systems and customary or traditional African laws regarding land ownership. In general, in most African countries, traditional land rights are obtained through inheritance with natural resources being considered communal and therefore accessible to all. Modern statutory laws, on the other hand, encourage state and private ownership. Emphasis under this system of law is also more on commercial production than on household production.[59] Therefore, when conflicts arise from the differences between these two systems of law, access to land definitely becomes problematic since under the two legal systems access to land is premised on different principles. Consequently, long-term negative impacts on the environment become inevitable, which further negatively affects the livelihood security of local communities. As a result, even further environmental degradation ensues.[60] Interestingly, even though local communities have the tradition of bushmeat conservation, there still exist local community management practices of forests resources. However, with the weakening of traditional access through state control, unsustainable practices such as the indiscriminate hunting of bushmeat have instead resulted in the weakening of local community management practices of forest resources.[61]

5.3.5.4 A way forward: the importance of indigenous values in conservation efforts

Cultural practices were successfully used by many traditional societies to maintain sustainable human communities before the period of industrial revolution and urbanization. Although bushmeat hunting may lead to depletion and extinction of many wildlife species, this might not have been the case had hunting been restricted only to indigenous Pygmies. However, with the commercialization and increased demand from urban areas, bushmeat hunting has become pervasive and unsustainable. Given that bushmeat hunting is so vital to the very existence of the Pygmies, it is imperative that conservation efforts take into consideration preserving such indigenous practices. Thus, in order to avert forest degradation, it is critical that states and governments recognize cultural practices and traditional beliefs and incorporate them into current national and international conservation plans and programs in ways that are sustainable for both the human communities and natural ecosystems.[62]

Unsuccessful biodiversity and conservation efforts have often been attributed to the failure of states and conservationists to consider local contexts and cultural values in the affected areas and how those factors would influence the outcomes of the initiative.[63] One relevant example comes from Hutton and Dickson, who use the bushmeat trade in Central Africa as evidence linking ecology with the economics of protected areas and the resources therein.[64] They argue that despite the conventional wisdom of many conservationists, who think that the consumptive use of resources must be either stopped or heavily regulated, the commercial use of resources is not in and of itself incompatible with conservation. Therefore, they argue, sustainable resource conservation can still be achieved even with some bushmeat hunting and consumption.

The question worth asking then is: Do indigenous peoples share concerns about the depletion and extinction of species and loss of biodiversity? Many experts argue that indigenous or local communities such as the Pygmies would oppose the delineation of "protected areas" if they were to limit certain traditional practices.[65] Conservation management projects that ignore cultural values frequently trigger conflicts and may limit and hinder the cooperation of local stakeholders.[66]

Therefore, for rural communities to have a vested interest in protecting biodiversity and managing the sustainable use of resources, it is important, where feasible and appropriate, to help them secure formal, legitimate, and exclusive rights to the benefits of wildlife and fish resources within lands and waters over which they have traditional claims. Bushmeat hunting must be regulated when it comes to exploitation for large-scale commercial purposes that disregard the impacts this has on biodiversity and conservation efforts. Ultimately, consultations with all stakeholders are key in order for all perspectives to be taken into consideration as well as to obtain the necessary engagement to engender conservation efforts while respecting and protecting indigenous rights and practices.

5.4 Case study: resource diversification and collaborative action by the Mbororo pastoralists in the North-West Region of Cameroon

5.4.1 Overview

Cameroon is located in West Africa, bordered by Nigeria to the west, Chad to the north-east, the Central African Republic to the east and Equatorial Guinea, Gabon, and the Republic of the Congo to the south. It has a population of about 23.5 million people, mostly Christian or Muslim, spread out over ten French- or English-speaking administrative regions. Mbororo pastoralists make up about 2.8 million of the population.

The country, which has been dubbed "Africa in miniature," is endowed with many natural resources, but poverty is still rife: 33 percent of Cameroonians live on less than $1.25 per day. The situation in the North-West Region, home to a large number of Mbororo settlers, is even more dire, as about 51 percent of the population there live below this level. The majority of the population in this area is made up of pastoralists and subsistent crop farmers.

The first Mbororos, a group mostly made up of Muslim pastoralists, came to Cameroon in the early part of the twentieth century, when the country was still under colonial rule. They migrated in droves to the North-West Region of the country, lured by its lush grazing lands and numerous salt springs. Led by Sabga, a fearless leader, these early Mbororo settlers appropriated for themselves a piece of land in the Bamenda highlands which became known as Sabga, as an initial dwelling place for the subsequent waves of Mbororo pastoralists who came to the area.

Under British colonial rule, the Mbororos were recognized as citizens. In fact, the British colonial masters encouraged Mbororo immigration to diversify the economy and augment state coffers.[67] However, the coexistence of the Mbororos with local communities was anything but peaceful; tensions between them and other settled groups were frequent, mostly revolving around land tenure.[68]

It is seemingly ambiguous that also under British colonial rule, the Native Land and Rights Ordinance (1927) was enacted to limit land ownership rights to only the inhabitants "whose parents belonged to a tribe indigenous to the Cameroons." This definition inherently excluded the Mbororo people due to their relatively recent arrival and nomadic lifestyle. With independence from Britain came little change, as the land ordinance of the independent West Cameroon was only slightly amended from the colonial Native Land and Rights Ordinance; the definition of a "Cameroonian" under the new law mirrored the definition of "indigenous" under colonial law. Suffice to say that in the newly independent state the Mbororo people were just as blocked from land ownership as they were under the colonial regime. Their only awarded right was a 25-year grazing permit given by the government of the West Cameroon state in a bid to secure revenue from them.

5.4.2 The Mbororos and agrobiodiversity

The Mbororos who settled in Cameroon are pastoralists whose livelihood has always depended on cattle rearing. They are the main producers and suppliers of beef and fresh milk for the Cameroonian economy and the pastoral livestock they keep accounts for 95 percent of cattle, goat, and sheep production in Cameroon. They are mostly concentrated in the North-West Region where they practice a semi-nomadic lifestyle: moving their animals down the valleys during the dry season in search of fresh pasture, crop residues, and water, and returning to the plateaus during the rainy season.

For decades, the Mbororos reared cattle with little, if any concern for soil preservation. Recently, however, they have delved into commendable agro-biodiversity efforts based on the use of biogas. Biogas is a type of cooking gas created through a rudimentary process called bio-digestion which converts cattle manure into cooking gas.

> Since the introduction of biogas for cooking, life is a lot easier [for the local communities] because meals can be cooked anytime due to the availability of energy from the biogas. Children go to school on time because they don't need to fetch wood again before going to school.[69]

To produce biogas, cattle manure is first stirred with a stick in an open container before being transferred to a tank where the bio-digestion naturally takes place.[70] This clean and smokeless alternative to wood has helped preserve plant life by reducing household dependence on wood as fuel for cooking.[71] Since 2013, at least 14 biogas plants have been built by the Mbororo Social and Cultural Organization (MBOSCUDA) to save energy and promote the conservation of natural resources.

Cattle manure is being prepared for bio-digestion (http://www.villageaid.org/about-us/projects/cameroon/#1443789260624-2-4)

The new practice of alliance farming, which is discussed in detail in the forthcoming paragraphs, is another strong biodiversity effort of the Mbororo people. Alliance farming involves the shared use of land for farming and grazing in rotation from season to season. Cattle are reared on the land after the harvest season to feed on the crop residue left from the harvest. As they graze on the land, the cattle leave behind manure and urine which serve as fertilizer for the land in preparation for the upcoming planting season. Thus, alliance farming continuously rehabilitates the soil and conserves its natural components.

5.4.3 Farmer–grazer conflicts in the North-West Region

In recent decades, the longstanding conflict between the Mbororos and indigenous subsistence farmers has been exacerbated by several factors. For one, there has been unprecedented and steady environmental degradation in the region, perhaps owing to overexploitation and poor conservation of the soil by both the farmers and grazers over time. Prolonged dry seasons and uncertain rainfall patterns in recent decades have further reduced the amount of viable land available for pastoral and farming activities, leading to an upswing in competition between the two groups over land and water use.

Over-extensive cattle grazing and negligence on the part of herdsmen also frequently lead to the destruction of crops. Some of the grazers and farmers do not maintain cattle-proof fences, leaving opportunities for cattle to wander into farmland to feed, thereby destroying crops. Moreover, when the cattle are being transported to transhumance areas and major cattle markets, they often destroy crops on their way, if not properly controlled. These destructive behaviors come at no small cost: the toll of destruction by cattle in just one incident could be valued up to 1 million fcfa ($1,700). In response, farmers whose crops have been destroyed by the cattle sometimes adopt very drastic retaliatory actions, such as killing or injuring the cattle.[72]

According to a local farmer:

> the pastoralists are very stubborn and don't like to take care of their animals. When the animals destroy crops few of them are willing to negotiate with us the farmers. The act occurs repeatedly because they are stubborn and don't care about food crop production. Some of them prefer to go and pay greater amount of money to the government administration instead of compensating the farmer whose crops have been destroyed.[73]

Moreover, population growth in the North-West Region and Cameroon in general has led to an increase in the demand for food, meat, and water. As the pressure on grazers and farmers to feed this growing production increases, so does the competition between the two groups over the use of the limited land and water resources to generate enough income required to meet their own basic needs.[74] As both farmers and grazers vie for the use of limited land, it is very common to see farmers complain about the encroachment of cattle into farmlands, while the pastoralists complain about the encroachment of farmers in grazing lands. This conflict is exacerbated by the Cameroonian government's failure to delineate specific land boundaries for grazing and farming activities in a way that meets the needs of the population, despite stipulations delineated in the Farmer/Grazer Act of 1978, which attempted to regulate farming and grazing activities.

Finally, some members of the Agro-Pastoral Commission and judiciary have vested, albeit corrupt, interests in the conflict between the pastoralists and farmers and contribute fuel to its continuing existence. To the unscrupulous members of the Agro-Pastoral Commission, the conflicts present good opportunities to collect bribes; while for those in the judiciary the conflicts serve as money-making opportunities from legal representation, even where it is not warranted.

On paper, Cameroon has well-developed laws to guide farmer–grazer relationships. According to the 1974 and 1976 land laws (Nos. 74-1, 74-2,

74-3, 76/156, 76/166, and 76/167), which were modified by Decree No. 2005/481 of 16 December 2005, five types of land exist in Cameroon:

- National Land
- Public Property of the State
- Private Property of the State
- Private personal land
- Land covered by final concession[75]

There are situations where some Divisional Officers (DOs) take it upon themselves to go into the field without being accompanied by the other members of the Agro-Pastoral Commission. There is the circumstance of a farmer whose crops, valued at 1 million fcfa ($1,700), were destroyed by cattle and a DO went to the field unaccompanied and took a cow from the grazer in question. He then bullied the farmer, caring little about the event because he had already collected a bribe.

On the other hand, lawyers come in to encourage grazers to file complaints because of their own financial interests, even when amicable settlements have been agreed upon. They give the impression that the Agro-Pastoral Commission is out to extort money through amicable settlement. Consider this example: A grazer's sheep destroyed the crops of eleven farmers and the farmers herded the cattle and sheep to the DO's office. After deliberating with the disputing parties, the grazer agreed to pay the farmers 712,900 fcfa ($1,200). When the date for settlement arrived, rather than give the farmers the agreed amount, the grazer brought a petition written by a lawyer. The matter was taken to the Senior District Officer, who examined the report of the Agro-Pastoral Commission that stated that the grazer had agreed to an amicable settlement. When asked why he could not settle the issue as earlier agreed rather than follow what the lawyer was vying for, the grazer had this to say: "When those lawyers write, does one read again?" Lawyers do not do a job like this free of charge. The grazer paid for the services and in significant sums of money.[76]

Land for use in farming and grazing mostly falls under the definition of National Land. Under these statutory provisions, National Land is managed by the Agro-Pastoral Commission, which was created by Decree No. 78/263 of July 3, 1978 (Farmer/Grazer Act 1978), establishing the terms and conditions for settling farmer–grazer disputes. The commission is headed by the DO of the local subdivision and is composed of a representative of the Divisional Service for Lands, a representative of the Divisional Service for Surveys, a representative of the Ministry of Agriculture, a representative of the village chief of the area in question, two notables, and a cattle grazer (known locally as an "Ardo").

The Agro-Pastoral Commission performs the following duties in its management of farmland:

- It allocates land for farming and grazing in rural areas based on the unique population and development needs therein (though this allocation and any subsequent modifications must be ratified by the regional governor).[77]
- It oversees the use of land allocated for farming and grazing, ensuring that established boundaries are well respected by both the farmers and grazers.
- It adjudicates on infractions between the aggrieved parties where said infractions are civil in nature (criminal offenses are considered crimes against the state and are handled only by the law courts).[78]
- It spells out guidelines for the use of mixed farming zones. These are transhumance zones: pieces of land not allocated permanently to anybody but used alternately by farmers and grazers on a seasonal basis. The Commission determines the period of the year when crop farming and grazing occur. (Alliance farming, an emerging trend among farmers, discussed in the later part of this chapter, makes use of mixed farming zones).

In addition to these laws and decrees, the country's constitution was revised in 1996 to establish freedom of movement and settlement for any of its citizenry within its territorial boundaries. Despite these robust-seeming laws, farmer–grazer conflicts have not abated, largely because of widespread corruption among the judiciary, law enforcement, and territorial administration that greatly impedes the equitable application of these laws. In fact, since 1982, there has been very little, if any, just allocation of rural land for farming or grazing purposes according to the economic and development needs of the region, as the scale of justice in farmer–grazer conflicts usually tilts in favor of the person with the highest amount to money to sway the corrupt officials. The Mbororos, who are mostly illiterate and often less aggressive, usually find themselves at a disadvantage when dealing with the government-appointed mediators. The blatant discrimination helps to aggravate the already strained relationship between the parties, forcing stakeholders to seek alternative strategies for conflict resolution.

5.5 *Alternative means of conflict resolution*

MBOSCUDA, in partnership with other international NGOs, such as the EU, Village, and Comic Relief, has dedicated much time and many resources to mitigate the incidence of conflict between the Mbororo pastoralists and subsistence farmers in the North-West Region of Cameroon through a five-year project called "In Search of Common Ground," introduced in 2013. This collaborative effort seeks to set up and encourage agricultural interventions (such as alliance farming) that can serve as a springboard for conflict resolution over the use of scarce resources. The following strategies have been adopted to achieve this aim:

- Dialogue platforms have been introduced by MBOSCUDA to facilitate cooperation between farmers and herders.
- Agricultural practices that promote shared use of resources, such as alliance farming, have also been introduced.
- Water-management strategies have been revised to ensure better access to clean and safe water.

5.5.1 *"Dialogue platforms"*

"Dialogue platforms" are community meetings facilitated by specially trained members of the grazing and farming communities where the members of the two groups can come up with amicable settlements to resolve problems, issues, and conflicts.[79] For instance, a farmer who has a complaint about the destruction of his or her crops by cattle can report the matter to the dialogue platform instead of taking up retaliatory (and often violent) action or seeking prolonged legal redress, as they would have done in the past. The meetings have also enhanced understanding and goodwill and improved longer-term relations between the farming and herding communities.

Dialogue platforms have been experimented with by MBOSCUDA and the Netherlands Development Organization (SNV) in lieu of the corrupt governmental Agro-Pastoral Commission and with reportedly higher success at conflict resolution. For example, between 2006 and 2010, dialogue platforms successfully reduced farmer–grazer conflicts in parts of the North-West Region by about 40 percent. Given that the dialogue platforms are preemptive in nature, they are considered a good opportunity to solve conflicts sustainably, since, in the forum, the principal parties in conflict meet to seek solutions to little squabbles before they escalate into full-blown conflict.

5.5.2 *Shared use of resources*

Cattle grazing on land previously used for farming (http://mboscuda.org/index.php/our-projects/alliance-farming/14-contradictions)

The success of the dialogue platforms has also paved the way for the successful sharing of resources between the farmers and grazers, as both parties are finally able to see their relationship as mutually beneficial: farmers provide food for cattle with crop residues after harvests, while grazers provide fertilizer to the farmers with cattle manure. Based on this symbiotic relationship, the two groups have realized they are able to share the land resources. From this realization, alliance farming was born.

Alliance farming is a partnership between a pastoralist and a subsistent crop farmer on shared use of resources such as land, water, and pasture. It was born out of a conflict-resolution dialogue between community-based farmer-herder groups, MBOSCUDA, and other NGOs over land use. Under the alliance farming system, the farmers make use of the land during the planting season. Their crops are fertilized by the manure left by the cattle that graze on croplands after harvest to consume the residual vegetation. After every harvest season, an area of land is fenced off for grazing. This post-harvest land contains tiny shoots of new plants which are highly nutritious for the cattle. Although the cattle only graze on the land for a few months, they leave the land in a more productive state for the coming season's crops, since it only takes about a month for the nutrients and minerals in the cattle urine and manure to fertilize the soil.

Since its inception in the region, alliance farming has been successful not only in deterring conflict and improving collaboration between the farmers and the grazers, but also in improving crop yield and cattle health. Due to its many benefits, alliance farming has steadily been gaining popularity among the local farmers and grazers. Within three years of its introduction, the number of grazers who adopted alliance farming increased by about 45 percent, and the number of farmers engaged in the practice increased by 33 percent. Moreover, 99 percent of the farmers involved report higher crop yields under alliance farming.[80]

5.5.3 *Water management*

In the past, conflict between the farmers and grazers has also stemmed from the lack of clean water availability in the local communities, which the farmers attributed to pollution from cattle bringing in water-borne diseases such as typhoid and diarrhea. The lack of clean water in the area meant that in the dry season women and children had to walk hours to the nearest water source. Consequently, the children were often late for school.

As a conflict-resolution strategy in its five-year project, MBOSCUDA, in partnership with its international collaborators has constructed water-catchment areas, storage tanks, and standing taps to deliver clean water to the communities. Fencing and planted trees act as barriers to water leakage and help keep cattle away, and the cows now use separate drinking troughs to avoid contamination. To ensure sustainability, water-management committees across various communities manage the conservation efforts and ensure the needs of the whole community are represented.

> So much has changed since Village Aid helped us complete the water project here. The children used to have to walk very far to collect water. They were late for school so their education was affected. Diarrhea was a big problem. We often had to fight to get water … The change in our lives is so marvelous that I have no words to express our happiness and gratitude about that has been done. You have changed lives, especially for women.[81]

These water projects have served as a unifying force between the farmers and grazers—as evidenced by a decline in conflicts over water shortages and pollution. The water-management projects have also succeeded in educating the farmers and grazers about the mutual benefits of the sustainable management of agro-pastoral resources.

5.5.4 *Conclusion*

The contribution of the Mbororo people to biodiversity efforts in Cameroon cannot be overemphasized, given their adoption of unique practices (such as alliance farming and biogas use) that have had far-reaching impacts on the conservation of the natural resources of the soil they utilize for their economic activity. Even more relevant to the discourse are their contributions to conflict resolution: the simple yet effective evidence-based strategies they have employed can easily be adopted to quell turbulence in other, similar jurisdictions. Despite this progress, much work remains to be done, as these praiseworthy efforts have only been introduced in select Mbororo communities.

5.6 Moving forward: stakeholder dialogues in policy formation

The foregoing discussion has shown three distinct practices with varying impacts on biodiversity. While the water-harvesting techniques of the Sahel region are received with much praise for their successes in soil conservation, they have faced some backlash from governments over the conversion of certain lands for agricultural use. In the Congo Basin, bushmeat hunting practices, central to the livelihoods of indigenous Pygmy tribes, have in recent years been exploited on a much larger scale for economic gain and thus faced widespread criticism by the international community for their contribution toward the extinction of wildlife species and the spread of diseases. Meanwhile, the relatively new conflict- resolution strategies of the Mbororos that have made a laudable impact on plant and animal life are yet to be wholeheartedly embraced and instituted in all their nomadic communities.

What, then, is the way forward in an era marked by relentless calls for the preservation of plant and animal life, and unique indigenous cultural inclinations of a people for future generations? A ban on the hunting and consumption of bushmeat, for example, would be an affront to the cultural heritage of the Pygmies as much as a mandate for a sedentary lifestyle would be to the Mbororos. Going forward, then, more dialogue is needed between policy-makers and indigenous groups in the formulation of policies that would successfully foster biodiversity efforts while still preserving the cultural heritage of the people.

5.7 Editors' note

This chapter highlights conservation efforts in the context of indigenous cultural inclinations in Africa and their effects on biodiversity. The authors Agbor and Elangwe provide an important account of specific examples and reflections for the future.

Illustrations of this chapter may continue with a dialogue about the wildlife-livestock interface, the relationship between agrobiodiversity and wildlife. In many African regions, bush-meat illustrates these intersections, where pathogen flow and genetic diversity provide a special discourse for regulation.

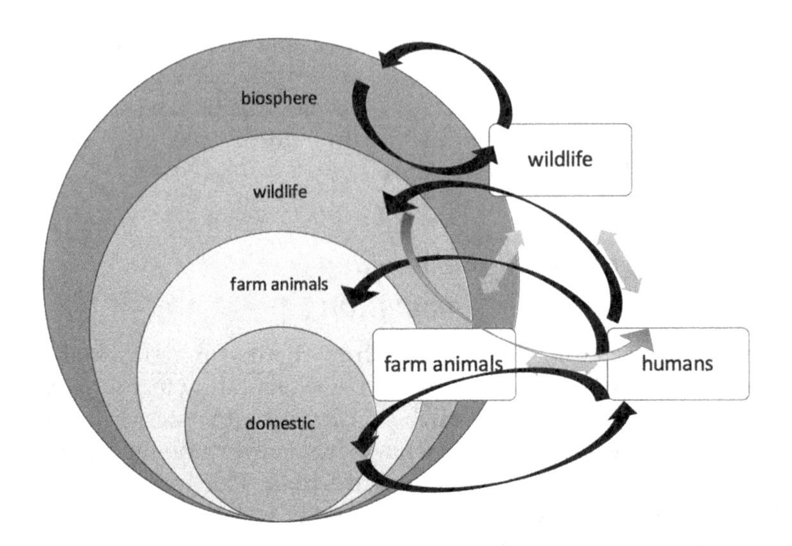

This figure illustrates that every realm, domestic, farms, wildlife and the larger biosphere, are effected and affect humans, farm animals and wildlife. The arrows illustrate pathogen flows but can be viewed as pathways for various forms of relationships, as described in the chapter.

Notes

1. "Sahelian Acacia Savanna," World Wildlife Fund (undated), www.worldwildlife.org/ecoregions/at0713.
2. "Sahel: $1.6 Billion Appeal to Address Widespread Humanitarian Crisis," United Nations Office for the Coordination of Humanitarian Affairs (2013), www.unocha.org/story/sahel-16-billion-appeal-address-widespread-humanitarian-crisis.
3. "Good Practices in Agricultural Water Management: Case Studies from Farmers Worldwide," United Nations Department of Economic and Social Affairs (2005), https://web.archive.org/web/20180417194705/http://www.un.org/esa/sustdev/csd/csd13/documents/bground_3.pdf.
4. Ezeanya-Esiobu, Chika, "Traditional Farming Practices for Enhanced Food Security," *Muslim Science* (September 2, 2014), http://muslim-science.com/traditional-farming-practices-enhanced-food-security/.
5. Kaboré, Daniel, and Chris Reij, "The Emergence and Spreading of an Improved Traditional Soil and Water Conservation Practice in Burkina Faso," Ifpri.org. (2004), www.ifpri.org/publication/emergence-and-spreading-improved-traditional-soil-and-water-conservation-practice.
6. Ezeanya-Esiobu, "Traditional Farming Practices."
7. Peyton, N., "Burkinabe Farmer Wins 'Alternative Nobel' for Drought-Fighting Technique," Reuters (September 24, 2018), www.reuters.com/article/us-niger-agriculture-award/burkinabe-farmer-wins-alternative-nobel-for-drought-fighting-technique-idUSKCN1M41ZV.

8. Reij, Chris, and Ann Waters-Bayer, *Farmer Innovation in Africa: A Source of Inspiration for Agricultural Development* (London: Routledge, 2001).
9. Spielman, David J., and Rajul Pandya-Lorch (eds), *Proven Successes in Agricultural Development* (Washington, DC: International Food Policy Research Institute, 2010), p. 166.
10. International Federation of Agricultural Producers, *Good Practices in Agricultural Water Management: Case Studies from Farmers Worldwide* (New York: United Nations Department of Economic and Social Affairs, 2005).
11. Ezeanya-Esiobu, "Traditional Farming Practices."
12. Kaboré and Reij, "Emergence."
13. Indigenous expert Chika Ezeanya-Esiobu, quoted in Perroni, E., "The Value of Traditional African Knowledge in Agriculture," foodtank (2017), https://foodtank.com/news/2017/11/traditional-african-agriculture/.
14. Ibid.
15. Wedum, J., Y. Doumbia, B. Sanogo, G. Dicko, and O. Cisse, "Rehabilitating Degraded Land: Zai in the Djenne Circle of Mali," *AGRIS: International Information System for the Agricultural Science and Technology* (1996), http://agris.fao.org/agris-search/search.do?recordID=GB9708660.
16. Spielman and Pandya-Lorch, *Proven Successes*, p. 54.
17. Sawadogo, H., F. Hien, A. Sohoro, and F. Kambou, "Pits for Trees: How Farmers in Semi-Arid Burkina Faso Increase and Diversify Plant Biomass," in Reij and Waters-Bayer, *Farmer Innovation in Africa*.
18. Reij and Waters-Bayer report that, by 1999, Ousséni Zoromé's fields had about 200 baobab trees. Ousseni Kindo, another successful Zai farmer, had persevered and cultivated about 100 baobab trees on his reclaimed land. Ibid.
19. Haggblade, S. and P. Hazell, *Successes in African Agriculture: Lessons for the Future* (Washington, DC: International Food Policy Research Institute, 2010), p. 292.
20. Ouedraogo, A., and H. Sawadogo, "Three Models of Extension by Farmer Innovators in Burkina Faso," in Reij and Waters-Bayer, *Farmer Innovation in Africa*.
21. Kaborè and Reij, "Emergence."
22. Sawadogo, H., F. Hien, A. Sohoro, and F. Kambou, "Pits for Trees: How Farmers in Semi-Arid Burkina Faso Increase and Diversify Plant Biomass," in Reij and Waters-Bayer, *Farmer Innovation in Africa*.
23. Ezeanya-Esiobu, "Traditional Farming Practices."
24. Fa, John E., Jesús Olivero, Miguel Angel Farfán, et al., "Differences between Pygmy and Non-Pygmy Hunting in Congo Basin Forests," *Plos One* 11(9) (2016), https://doi.org/10.1371/journal.pone.0161703.
25. Ibid.
26. Ibid.
27. Friant, Sagan, Sarah B. Paige, and Tony L. Goldberg, "Drivers of Bushmeat Hunting and Perceptions of Zoonoses in Nigerian Hunting Communities," *PLOS Neglected Tropical Diseases* 9(5) (2015), https://doi.org/10.1371/journal.pntd.0003792.
28. Ibid.
29. Ibid.
30. Ibid.

31. Ibid.
32. "Indigenous Peoples in Cameroon," IWGIA (September 12, 2011), www.iwgia.org/en/cameroon/743-indigenous-peoples-in-cameroon.
33. "World Directory of Minorities and Indigenous Peoples: Gabon" (2007), Refworld, www.refworld.org/docid/4954ce3323.html.
34. Ibid.
35. Food and Agriculture Organisation of the United Nations, "Illegal Hunting and the Bush-Meat Trade in Savanna Africa: Drivers, Impacts and Solutions to Address the Problem" (2015), www.fao.org/3/a-bc609e.pdf.
36. Nasi, R., A. Taber, and N. Van Vliet, "Empty Forests, Empty Stomachs? Bushmeat and Livelihoods in the Congo and Amazon Basins," *International Forestry Review* 13(3) (2011): 355–368, https://doi.org/10.1505/146554811798293872.
37. Ibid.
38. Poulsen, J. R., C. J. Clark, G. Mavah, and P. W. Elkan, "Bushmeat Supply and Consumption in a Tropical Logging Concession in Northern Congo," *Conservation Biology* 23(6) (2009): 1597–1608. doi: 10.1111/j.1523-1739.2009.01251.x.
39. Ordaz-Németh, Isabel, Mimi Arandjelovic, Lukas Boesch et al., "The Socio-Economic Drivers of Bushmeat Consumption during the West African Ebola Crisis," *PLOS Neglected Tropical Diseases* 11(3) (2017), https://journals.plos.org/plosntds/article?id=10.1371/journal.pntd.0005450.
40. Enuoh, O., and E. F. Bisong, "Biodiversity Conservation and Commercial Bushmeat Hunting Challenges in African Parks and Protected Areas: A Critical Review and Synthesis of the Literature," *Research on Humanities and Social Sciences* 4(15) (2014): 39–57.
41. Retrived May 10, 2018 from URL: http://www.ibtimes.com.au/bushmeat-urban-households-threat-amazon-wildlife-1417670
42. Fa, John E., Sarah Seymour, Jef Dupain et al., "Getting to Grips with the Magnitude of Exploitation: Bushmeat in the Cross–Sanaga Rivers Region, Nigeria and Cameroon," *Biological Conservation* 129(4) (2006): 497–510, doi: 10.1016/j.biocon.2005.11.031.
43. Chaber, Anne-Lise, Sophie Allebone-Webb, Yves Lignereux et al., "The Scale of Illegal Meat Importation from Africa to Europe via Paris," *Conservation Letters* 3(5) (2010): 317–321, doi:10.1111/j.1755-263x.2010.00121.x.
44. Jones, Mark, "Is Africa's Wildlife Being Eaten to Extinction?" *BBC News* (August 3, 2010), http://news.bbc.co.uk/2/hi/science/nature/8877062.stm.
45. "Bringing Animal Products into the United States," Centers for Disease Control and Prevention, www.cdc.gov/importation/animal-products.html.
46. "Get Ready for the Next HIV and Ebola: Experts Warn Illegal Bushmeat Could Cause a New Food Pandemic," *MailOnline* (September 30, 2016), www.dailymail.co.uk/sciencetech/article-3816514/Get-ready-HIV-Ebola-Experts-warn-BUSHMEAT-cause-new-food-pandemic.html.
47. Friant, Paige, and Goldberg, "Drivers of Bushmeat Hunting."
48. Fa, et al., "Differences between Pygmy and Non-Pygmy Hunting"; see also Bennet, E., H. Eves, J. G. Robinson, and D. Wilkie, "Why Is Eating Bushmeat a Biodiversity Crisis," *Conservation in Practice* 3(2) (2002): 28.
49. Fa, et al. "Differences between Pygmy and Non-Pygmy Hunting."
50. Ibid.
51. Friant, Paige, and Goldberg, "Drivers of Bushmeat Hunting."

52. "World Health Organization, Situation Report" (June 10, 2016), http://apps.who.int/iris/bitstream/10665/208883/1/ebolasitrep_10Jun2016_eng.pdf?ua=1.
53. Judson, Seth D., Robert Fischer, Andrew Judson, and Vincent J. Munster, "Ecological Contexts of Index Cases and Spillover Events of Different Ebolaviruses," *PLOS Pathogens* 12(8) (2016), https://journals.plos.org/plospathogens/article?id=10.1371/journal.ppat.1005780.
54. Saez, A. Mari, Sabrina Weiss, Kathrin Nowak et al., "Investigating the Zoonotic Origin of the West African Ebola Epidemic," *EMBO Molecular Medicine* 7(1) (2014): 17–23, www.ncbi.nlm.nih.gov/pubmed/25550396; see also, Alexander, Kathleen A., Claire E. Sanderson, Madav Marathe et al., "What Factors Might Have Led to the Emergence of Ebola in West Africa?" *PLOS Neglected Tropical Diseases* 9(6) (2015), https://journals.plos.org/plosntds/article?id=10.1371/journal.pntd.0003652.
55. Food and Agriculture Organisation of the United Nations, "Illegal Hunting."
56. Ibid.
57. USAID, "Central Africa Regional Program for the Environment (CARPE) Regional Development Cooperation Strategy 2012–2020" (2012), www.usaid.gov/central-africa-regional/cdcs.
58. Ibid.
59. Ibid.
60. Ibid.
61. Ibid.
62. Asante, Eric Appau, Stephen Ababio, and Kwadwo Boakye Boadu, "The Use of Indigenous Cultural Practices by the Ashantis for the Conservation of Forests in Ghana," *SAGE Open* 7(1) (2017), https://journals.sagepub.com/doi/full/10.1177/2158244016687611.
63. Mbile, P., M. Vabi, M. Meboka, D. Okon, et al., "Linking Management and Livelihood in Environmental Conservation: Case of the Korup National Park Cameroon," *Journal of Environmental Management* 76(1) (2005): 1–13, http://europepmc.org/abstract/med/15854732.
64. Hutton, Jon, and Barney Dickson, "Conservation out of Exploitation: A Silk Purse from a Sow's Ear?" in John D. Reynolds, Georgina M. Mace, Kent H. Redford, and John G. Robinson (eds), *Conservation of Exploited Species* (Cambridge, UK: Cambridge University Press, 2002), pp. 440–461.
65. Enuoh and Bisong, "Biodiversity Conservation."
66. Muhumuza, Moses, and Kevin Balkwill, "Factors Affecting the Success of Conserving Biodiversity in National Parks: A Review of Case Studies from Africa," *International Journal of Biodiversity* (2013): 1–20, www.hindawi.com/journals/ijbd/2013/798101/.
67. Pelican, Michaela, "Mbororo Claims to Regional Citizenship and Minority Status in North-West Cameroon," *Africa* 78(4) (2008): 540–560, www.cambridge.org/core/journals/africa/article/mbororo-claims-to-regional-citizenship-and-minority-status-in-north-west-cameroon/7BC1F581E32ACD4993E23FBF536A4AC2.
68. Asong Tellen, Valentine, Juliana Anchang, and Martin Shu, "Conflicts Over Land and Pasture in North-West Cameroon: Listening to the Voices of Farmers and Grazers," in "In Search of Common Ground for Farmer–Grazer

Conflicts in the North West Region of Cameroon," Pan African Institute for Development—West Africa (PAID–WA) (April 27, 2016), www.paidafrica.org/paidwa/images/data/Qualitative_DREW_FINAL.pdf.

69. "Transforming Conflicts over Agro-Pastoral Resources into Mutually Beneficial Alliances," *Break Through Magazine, Volume III* (August 2015–July 2016), www.mboscuda.org/attachments/article/48/Annual%20Magazine-Break%20Through%203rd%20Edition.pdf.
70. "Cameroon, Village Aid," https://web.archive.org/web/20180320142128/www.villageaid.org/about-us/projects/cameroon/.
71. "Transforming Conflicts over Agro-Pastoral Resources."
72. Nchida, Valentine, Che Marcellus, Precillia Ijang Tata Ngome et al., "'In Search of Common Ground' for Farmer–Grazer Conflicts in the North-West Region of Cameroon: Midterm Evaluation Report," Institute of Agricultural Research for Development (IRAD), Cameroon (August 2016), www.researchgate.net/publication/308200169_In_Search_of_Common_Ground_for_Farmer_Grazer_Conflicts_in_the_North_West_Region_of_Cameroon_Midterm_Evaluation_Report.
73. Ibid., p. 18.
74. Ibid.
75. "Summary of Laws/Regulations Governing Pastoral Livestock in Cameroon," *MBOSCUDA* (October 2009), www.mboscuda.org/attachments/article/29/Laws Pastoral Livestock.pdf.
76. Nchida et al., "In Search of Common Ground," pp. 17–18.
77. "Summary of Laws/Regulations Governing Pastoral Livestock in Cameroon."
78. Ibid.
79. Nchida et al., "In Search of Common Ground," p. 24.
80. Nchida et al., "In Search of Common Ground."
81. A villager speaks in praise of water management efforts; see "Cameroon, Village Aid."

chapter six

Inclusive participation of persons with disabilities in aspects of agroecology and agrobiodiversity

Ivan K. Mugabi

Contents

6.1 Introduction

This chapter is related to the broader themes of this book, because of its fundamental contribution in correlating socio-legal interactions that subsist in valuing the voices of persons with disabilities in developing debates around agrobiodiversity and agroecology. This chapter demonstrates the attributes of considering disability in these global discussions that constitute one of the currently developing areas of public international law. Before proceeding, the chapter will use a wide range of literature on the interconnectivity of environmental rights[1] and land ownership rights, correlating with approaches of accessibility and inclusiveness,[2] in as far as the protection of disability rights is concerned.[3] That interconnectivity would promote inclusion of persons with disabilities in developing policies or practices of agrobiodiversity and agroecology relating to food safety and food security.[4] Therefore, useful recommendations shall be highlighted that might encourage the inclusiveness of persons with disabilities whose participation appears traditionally overlooked in debates shaping matters of agrobiodiversity and agroecology. Hence the first section relies on protocols of international law in rationalizing the inclusive participation of persons with disabilities when shaping laws or future policies on agrobiodiversity and agroecology. The final section elucidates on ways in which inclusive participation could be attained in a more practical sense. That section will consider means and methods of attaining inclusive participation.

6.2 Rationalizing the cause for a disability-inclusive participation

This section is designed to give an account of the meritorious attributes for supporting an approach of inclusiveness to the ideas and participation of persons with disabilities in developing policies or practices on agro-ecology and agrobiodiversity. At this stage, participative approaches that are structured in ways that seek to ensure effective inclusiveness of persons with disabilities might allude to Ludivine's idea of food democracy.[5] Perhaps it might be reasonable to commence this analysis with some sort of academic caveat that the reasons explained in the subsequent subsections are far from being exhaustive or conclusive. These reasons nonetheless give a sufficient justification for considering the involvement of persons with disabilities in the developing of laws and policies on agroecology and agrobiodiversity, in both domestic and international contexts. It must be recognized that although great strides have been made in according respect and special consideration to aspects of disability in most components of international agendas,[6] in terms of goals of scientific agendas,[7] persons with disabilities can still be classed as part of those groups whose opinions are more often than not afforded less consideration. It is therefore paramount to highlight the likely disadvantages that could arise in cases of affording persons with disabilities lower significance when structuring scientific agendas that affect their lives in the same way they affect able-bodied individuals. Of course, as has already been noted, in the context of this chapter, the structuring of those scientific agendas is with particular reference to addressing the contemporary matters of agroecology and agrobiodiversity.[8]

6.3 Persons with disabilities as consumers of food products in the production line

6.3.1 Social-care service providers and the extent of disability choices for dietary sources

Consumer rights in terms of food safety and food security should be non-negotiable and non-discriminatory. It must be recalled that persons with disabilities constitute more than 1 billion of the world's population, based on World Bank statistics, and the increasing population of persons with disabilities has impacted trends of food production for some time. For example, Schutter has analyzed the dramatic rise in the population in the 1960s.[9] Such population dynamics must be viewed in light of globally increasing numbers of persons with disabilities. In the same article, Schutter mentions the work of Ehrlich[10] and the relationship between processes of urbanization and agroecological interactions with food production.[11] In that regard, it might be worthwhile highlighting that, conventionally, limited consideration has been afforded to persons with disabilities in ensuring that their views, voices, and interests are considered in the course of making decisions concerning the nature of agroecological interactions comprising the supply chain of food production.[12]

Perhaps the globally increasing population of persons with disabilities might justify the importance of ensuring that more attention is given to this population so as to enhance the attention afforded this marginalized group while streamlining international agendas. This supposition assumes that persons with disabilities comprise the world's fastest growing population category. In some countries, the trend of detaining persons with mental disabilities in institutions is increasing the likelihood of portraying their needs as less important with regard to dietary aspects relating to food sources and food quality. That tendency is often worsened by the fact that a large number of them are confined in care homes,[13] which restricts their opportunities to participate in making food choices. Most private and public entities are keen to devise ways to ensure that the right to choose whether food should be secured from organic or inorganic suppliers. It is imperative to reiterate that having a clear understanding of the arrangements used in the social care industries is important in comprehending the external drivers accounting for ignoring the views of persons with disabilities when framing debates on matters of agroecology and agrobiodiversity. It must be understood that in recent decades the presence of disability rights have in many respects adopted aspects of independent living and revolutionized the importance of considering persons with disabilities while framing solutions to contemporary problems stemming from poor agroecology and agrobiodiversity practices.

Disability-enabling rights are instrumentally vital in appreciating the indivisibility of fundamental rights and freedoms. In other words, the right of persons with disabilities to participate in consultation forums that are concerned with future policies on agroecology and agrobiodiversity is partly dependent on the willingness of state and non-state actors to reach out to this population by promoting the availability and accessibility of consultative forums.[14]

Here it appears imperative to quote the provisions of the UN Convention on Rights of Persons with Disabilities regarding its perception of the disability-oriented view of the relationship between communication and accessibility:

> 'Communication' includes languages, display of text, Braille, tactile communication, large print, accessible multimedia as well as written, audio,

> plain-language, human-reader and augmentative and alternative modes, means and formats of communication, including accessible information and communication technology.[15]

This relies on the argument that protecting the inclusiveness of a right to a clean environment is indivisibly interdependent on promoting the accessibility of information concerning policies of agroecology and agrobiodiversity to persons with disabilities.[16] Such accessibility would increase the availability and capability of persons with disabilities to understand relevant information that remains a prerequisite to taking part in numerous discussions that are shaping policies on agroecology and agrobiodiversity.[17] As such, for purposes of inclusiveness, persons with disabilities need to be represented in prospective debates on agroecology and agrobiodiversity and means provided to make them accessible. The right to access information on policies of agroecology and agrobiodiversity remains key in ensuring and encouraging the inclusive participation of persons with disabilities in these aspects. That point of indivisibility shall be advanced further in subsequent sections of this chapter in order to demonstrate its role to the inclusion of persons with disabilities in aspects on agroecology and agrobiodiversity at national and international levels, thereby increasing the possibility of updating on aspects of food security and food safety, especially in terms of how these could affect interactions with environmental rights and their enjoyment of land uses.

It is vital to recognize that persons with disabilities are also customers for goods produced and manufactured in the food supply chain.[18] Accordingly, considering their involvement is a key component to understanding international and national policies on aspects of agroecology and agrobiodiversity. So it is imperative to acknowledge how persons with disabilities are part of the stakeholders in all sectors of agroecology and agrobiodiversity. Such a perspective would be key in and that increasing their visibility and boosting their ability to enjoy the benefits and effectiveness of agroecology and agrobiodiversity.

6.4 The United Nations Sustainable Development Goals demand for inclusion in all disciplines

The transformation of disability from a medical- to a social rights-based model has made international organizations and organs of the United Nations recognize the relevance of including persons with disabilities in consultation forums discussing a wide range of matters.[19] Although in some cases there are links between disabilities and obesity, there is still room for ameliorating the cause and effect and ensuring food security. That demonstration has appeared relatively limited in relation to the work of Aguirre.[20] Although Aguirre's work is unrelated to Sustainable

Developments Goals, it must be appreciated that this work examines the European Union's Common Agricultural Policy (CAP).[21] CAP may have been impacted by or influenced the goals in relation to ensuring agricultural policies are central to fostering food safety and food security with a primary incentive of ensuring inclusive policies.

6.4.1 *Human rights-based argument as rationality for inclusive participation*

With regard to promoting the rights of owning land and rights over land without disability-based discrimination, the idea of promoting and protecting the rights of land ownership could be embodied within the concept of ensuring equal property ownership.[22] That obligation is stipulated within the CRPD under Article 12(5): "Subject to the provisions of this article, State Parties shall take all appropriate and effective measures to ensure the equal right of persons with disabilities to own or inherit property."[23]

Bearing in mind that persons with disabilities comprise part of the potential landowners, it sounds logical to ensure that they are kept informed and aware of environmental information and scientific decisions that could directly or indirectly affect the fertility or productivity of their agricultural land. As such it becomes apparent that information on the engineering of climate-resilient plants ought to be communicated in ways that can be accessed among all persons with disabilities. That argument demonstrates a positive gesture of promoting and protecting rights that are related to owning and using land as an example of corporeal and incorporeal property.

However, the dissemination of relevant information such as that related to the engineering of climate-resilient breeds or crops is often delivered in ways that continue to exclude persons with disabilities from the advancement of scientific debates.[24] And yet, it would in the best interest of landowners, irrespective of whether they have hearing, visual, or intellectual impairments to be involved and informed of what constitutes climate change,[25] what are genetically modified plants,[26] or when and how engineered plants in an adjacent field might affect their land. Persons with disabilities are hardly represented in institutions of the United Nations, such as the United Nations Environment Programme (UNEP), which is wholly comprised of members from the World Meteorological Organization (WMO) and the Intergovernmental Panel on Climate Change (IPCC).[27] The possibility of giving limited space to a diverse assemblage, including persons with disabilities, to discuss matters relevant to climate change has been mooted,[28] and consequently voices of this marginalized group barely played a role in the preliminary arrangements for drafting the modern day United Nations Framework

Convention on Climate Change (UNFCCC).[29] The point of argument being made in this regard relates to suggesting that discussions and consultative forums that might affect the productivity of land by impacting the environment and food supply ought to be arranged in ways that are accessible to landowners with disabilities, since property ownership is comprised of additional ancillary rights that might remain unprotected in the absence of a disability-inclusive approach.

6.5 The interconnectivity of rights in the disability context

The idea of the interconnected nature of rights and the indivisibility of those rights is neither a novel nor a peculiar concept to persons with disabilities. Pierri explores a possible correlation ideas of rights in relation to aspects of agrobiodiversity.[30] Some further analysis could also be advanced in demonstrating the overlaps attributable to the interrelationships between health rights, the rights to accessing clean water, and the right to a healthy environment.[31] In a similar ethos, Kanter identifies the importance of extending the understanding of the indivisibility concept to the perception of disability-related issues,[32] as opposed to Pierri, whose interest is centered on rights and correlate with issues of agrobiodiversity.[33] It is apparent that scholarly ideas that assist in linking the indivisibility of disability rights to matters of agroecology and agrobiodiversity are relatively scanty and certainly still seem largely lacking.

This raises the issue of CRPD rights in terms of accessibility to communicated information and the mobility of persons with disabilities to places where discussions related to matters of agrobiodiversity and agroecology are conducted. Information concerning intellectual property rights regarding the genetic modification and eventual ownership of plant cells[34] should not be communicated in ways that exclude persons with disabilities.[35] It could be argued that such information has in the past neither been made available nor made accessible to persons with disabilities. Arguably, there is a duty to ensure that efforts are made to convey information concerning matters of agrobiodiversity and agroecology in ways that make them accessible and available to persons with disabilities.

6.5.1 Environmental rights of persons with disabilities and resource exploitation

The concept of environmental rights is interconnected to health rights and ultimately linkable to the right to life. It should be borne in mind that other property-related rights, such as the right of ownership and specifically the use of land, could be negatively impacted in the absence of protection of environmental-related rights. It can also be argued that the ability to use land for food production to enhance food security and food safety could be positively or negatively impacted by the extent of protecting and respecting environmental rights.[36] For instance, if the level of protection associated with the importance of respecting environmental rights is high, then there is a greater possibility that actions which could impact the environment negatively will face significant resistance from all people, including persons with disabilities. By way of contrast, if the degree or level of protection associated with the importance of respecting environmental rights is low, then there is a likelihood that actions which negatively impact the environment will thereby be faced with insignificant resistance, those actions far from being discouraged. That presumption holds credibility if the influence of drivers from other economic, political, and social institutions is presumed to be a constant rather than a variable.

In view of the above environmental suppositions, persons with disabilities must be encouraged to participate in consultation forums for framing policies and laws that regulate national and international sectors of agroecology and agrobiodiversity.[37] Their participation in the framing such policies remains long overdue despite the likelihood of activities carried out by industrial operations associated with acts of agrobiodiversity to positively or negatively impact the extent of respect for the environmental rights of all persons, including those with disabilities.[38] That perhaps explains why literature on the concepts of agroecology and agrobiodiversity is often linked to contemporary debates on matters of global warming.[39] Eventually, the cost of ignoring the importance of promoting environmental rights[40] might lead to natural disasters that have far-reaching consequences on the livelihood of persons with disabilities, especially if compared with other vulnerable groups.[41] For example, persons with disabilities are more likely to be left behind in cases of floods, which might be caused by the activities of mankind compromising the importance of environmental protection.[42] Similarly, in regions that face food crises and the consequent famines, persons with disabilities are likely to be impacted more compared with able-bodied people.[43]

Bearing in mind that famine leads to hunger, the 1966 report of the Food and Agriculture Organization of the United Nations demonstrated the connection between efforts to eradicate hunger and tendencies of direct or indirect acts of disability-related discriminative eradication practices.[44] In this context it is perhaps suitable to quote the propositions asserted by the report:

> Fighting hunger in the world means fighting to feed all hungry people. The rural disabled are a particularly vulnerable group: they and their households

> tend to be the poorest of the poor; they have special needs, but are often marginalized and overlooked in development interventions.[45]

6.5.2 *Indiscriminative distribution of food safety and food security information*

The notion of equality is established under the Convention of Rights of Persons with Disabilities.[46] However, in the same document, its preamble admits that "despite these various instruments and undertakings, persons with disabilities continue to face barriers in their participation as equal members of society."[47]

The same preamble provides for:

> a comprehensive and integral international convention to promote and protect the rights and dignity of persons with disabilities will make a significant contribution to redressing the profound social disadvantage of persons with disabilities and promote their participation in the civil, political, economic, social and cultural spheres with equal opportunities, in both developing and developed countries.[48]

In terms of exploring the links between equality and ideas on the equitable distribution of food safety and food security information,[49] first and foremost, FAO has acknowledged the importance of recognizing a historical problem of the excluding of persons with disabilities in conveying information and delivering of programs related to aspects of the supply chain for food production.[50] It is imperative to underline that food production has some correlations with aspects of food safety and food security. The connection between food safety, food security, and food production means that attaining food safety and security involves seeing food processing as the starting point at which other stages such as bioecological productivity or conservations methods take place.[51] In response to the FAO's identification of the aforementioned historical problem, this UN department is making progress in integrating farmers with disabilities in its planning strategies.[52] It is worth crediting the FAO for its inclusiveness of farmers with disabilities in the framing of its global strategies. The FAO has also gone an extra mile in encouraging interactive dialogue about inclusive good practice among different states.[53] For example, in July 2010, 20 participants representing persons with disabilities from different states attended a global forum on food security and nutrition.[54] Countries that participated included Italy,[55] Nepal,[56] Kenya,[57] Uganda,[58] Canada,[59] the Netherlands, Germany,[60] South Africa,[61] Nigeria,[62] Mexico,[63] and Austria.[64] During the forum, the participants from these countries that comprised both developed and developing states exchanged ideas on what has seemed to work in their respective countries. The theme was aimed at identifying how their countries, regions, and institutions have taken measures to facilitate the inclusion of persons with disabilities into policies and programs related to food security and nutrition.[65] Subsequently, the participants discussed ways of convincing policymakers that without the involvement of organizations representing persons with disabilities and the mainstreaming of disability issues on Millennium Development Goals (MDGs) and Right to Food agendas these goals and rights are very unlikely ever to be achieved.[66]

In general terms, there is now a possibility that farmers with disabilities will be afforded well-deserved consideration in accessing FAO farming and disability-inclusive reports.[67] However, there is still limited evidence to suggest that the FAO shall sustain a continued practice of prioritizing the best interests of persons with disabilities. It is clear that there is still room for addressing the exclusion of persons with disabilities in relation matters of food safety and agro-based approaches that enhancing a greater compliance with inclusive practices that fit the broader definition of disability in order to reflect compatibility with that of the CRPD. To this effect, CRPD Article 1 stipulates that:

> Persons with disabilities include those who have long-term physical, mental, intellectual or sensory impairments which in interaction with various barriers may hinder their full and effective participation in society on an equal basis with others.[68]

It could be argued that any tendency that might result in excluding a specific group of individuals with disabilities, either intentionally or coincidentally, is likely to be divisive in nature, and that persons with disabilities are used to encountering strong attitudinal barriers in contemporary environments. In other words, applying the concept of inclusion for persons with disabilities must neither over emphasize the disability nor afford preferential attention to the impairment than appreciating that these are farmers with disabilities, as asserted in the mission statement of FAO: "Farmers with disabilities should be considered farmers first and people with disabilities second."[69]

But the above proposed perception rethinking about persons with disabilities as farmers with disabilities could afford them inclusive treatment compared the inclusiveness that they tend to be afforded in present consultations on agricultural debates. That perspective might be unjust for a number of reasons that will be explained further in the following subsections.

(a) It is contrary to the principle of inclusiveness.

The idea of promising farmers with disabilities special privileges seems contrary to the principle of incorporating norms of equality that embody in them attributes of inclusiveness by encouraging a perspective similar to that of equal treatment as enshrined in the various United Nations Human Rights Instruments. In that regard, it is imperative to contend that there is no rationale for the FAO to start redrawing the contents of the international definition of disability. Of course, in principle, it is wrong to support the idea of favoritism and perhaps imply that persons with disabilities who are not farmers or who identify with other occupations must be afforded comparatively less consideration than that given to farmers with disabilities.[70] Sadly, it could be asserted that in practice such a perspective seems to suggest that persons with disabilities in other professions might have to consider joining the agricultural sector for purposes of enjoying the same level of disability rights as those of farmers with disabilities that are alleged to comprise a distinctive category. The ideology of concentrating only on farmers,[71] while excluding other persons with disabilities that might be buyers with disabilities, or potential agricultural researchers and students with disabilities, sounds hardly empowering, especially to persons that belong to those other categories.[72] Of course, there may be practical reasons for the FAO's approach, but whatever those reasons might be, they are undoubtedly conflicting with fundamental norms of equalizing opportunities for persons with disabilities,[73] and these are entrenched in human rights agendas and supported by disability rights scholars.[74]

(b) The approach of the FAO seems more subjective and far from being objective in its conceptualization of disability.

The criterion for determining who is entitled to special consideration on grounds of their disability must be based on the fact that such a person meets the definition of disability as provided in the relevant international or domestic instrument.[75] In an international context, it might suffice to assert the same disability rights without discrimination on any grounds as long as the person has a long-term mental, physical, intellectual, or sensory impairment that hinders his or her daily interactions or effective participation in daily agricultural activities on an equal basis with others.[76] In other words, the test of the CRPD sounds inclined to objectivity rather than opening room for different actors and stakeholders in various agricultural departments to assume their subjectivity criterion. The subjective argument versus the objective test is relevant in terms of deciding which individuals must be considered by sectors of agroecology and agrobiodiversity. A subjective approach is likely to permit personal preferences of stakeholders to take precedence in determining who can be afforded consideration in policy initiatives,[77] as opposed to a legal certainty more likely to be associated with the objective test.[78]

> Mentally disabled trainees will need special guidance and direction. They can be very clever and (in some cases they have been established to) make very good mushroom farmers. Their attention to minute details sometimes makes them better in the maintenance of their mushroom farm and therefore, generate yields that are sometimes higher than non-mentally disabled people.[79]

It is highly probable that the fractionalizing of what is already a minority group by using a new criterion might easily lead to uncertainty and fractioning of persons with disabilities in terms of equality before the law. The analogy of separating and treating farmers with disabilities as kings while treating anyone else through a minimalist approach is synonymous with the idea of separating oranges from apples to afford better storage to another category of fruit. In other areas, such subjectivity as to who to permit to pick and choose could be divisive hence bound to raise moral, ethical, and human rights dilemmas.

(c) Distributive equity rather than social justice

The issue of equitable justice is another fundamental aspect that could cause more problems with the idea of putting farmers with disabilities first above other persons with disabilities.[80] Excluding factions of marginalized groups from special consideration is likely to conflict with an ideal approach for proper determinants of inclusive social justice.[81] Examples of those determinants include: addressing education inequalities, responding to food inequalities,[82] and health and income inequalities,[83] thereby subsequently structuring sectors of agroecology and agrobiodiversity in such a manner that they contribute to encouraging determinants of social justice indiscriminately by supporting persons with disabilities. Much as the previously mentioned sectors can support relevant determinants for displacing barriers of social exclusion while alleviating problem of public policy.

The above-mentioned barriers are characterized by overlooking the need to encourage practices that symbolize the inclusive communication of such information to persons with disabilities. This state of affairs appears to conflict with aspirations that are enshrined in the preamble of the aforementioned Convention.

In the above context, disseminating information on matters related to or affecting food security and food security must be equitably distributed by acknowledging the value of safeguarding persons with disabilities from exclusionary tendencies and attitudes. Those tendencies are often contrary to the goal of integrating persons with disabilities in different matters of public life.

6.6 Means and methods of enhancing inclusive participation

6.6.1 Accessibility of food production places and spaces as points of exclusion

6.6.1.1 Mobility aids to food places and spaces: accessibility of animal slaughterhouses, in context

It is imperative to appreciate that better inclusiveness would entail undertaking positive measures to make animal slaughterhouses and other food-packing places more accessible to persons with intellectual, sensory, physical, and, where practicable, those with mental disabilities.[84] In that regard, it is imperative to highlight that building materials and accessibility requirements of slaughterers may differ between developed and developing countries due to the variance in political, socioeconomic and institutional structures underlying the supply chain of meat production.[85] In this context, it must be appreciated that socioeconomic and political institutions through which products from the agroecology and agrobiodiversity industies are produced tend to differ in the context of developed countries from those in developed countries.[86] For example, in developing countries, challenges to the accessibility of slaughterhouses to persons with disabilities ought to be considered light of the varying availabilty and affordability of building materials in rural and urban settings.[87] The increasing numbers of persons with disabilities is a key justification for ensuring that government regulation on agricultural design becomes more inclusive, with washrooms that are accessible to accessibility and mobility devices of persons with disabilities such as agricultural structures with provisions for wheel chairs.[88] It is important to realize that most animal slaughterhouses are scarcely accessible, which hinders persons with disabilities from either reaching or working in such places. This also makes it is hard for persons with disabilities to fit into such spaces either as workers or customers.

In developed countries, there is significant disconnection between places such as slaughterhouses for processing food and locations for storing and selling food items such as warehouses and markets. In this case it is highly probable that a customer-based concept justifies considering the design of modern structures to accommodate all employees, including persons with disabilities in a slaughterhouse.[89] The employability of persons with disabilities in those places could be advanced by the creation of disability-friendly spaces,[90] for example, disability mobility-supportive facilities in relevant work stations. In both developed and developing countries, there are cases where the best interests of persons with disabilities are better met by domestic suppliers of organic and inorganic food and animal products who implement labeling and packing measures that are inclusive of persons with disabilities. Thus the conventional approach would include persons with disabilities especially in cases where fresh meat products are preferred to ready-made products that are often frozen.

6.6.1.2 Accessibility of specific farming localities

In this context, the inclusiveness of persons with disabilities is seen in relation to farming sites where various agricultural activities are conducted. Laws regulating agriculture-related land uses are often unclear in terms of specifying the different places that should be made inclusive for persons with disabilities, bearing in mind that agroecology- and agrobiodiversity-related products are likely to be obtained from various agricultural processes, some of which are worthwhile outlining herein. For instance, floriculture, which is concerned with the art and science of growing as well as processing flowers;[91] horticulture, which refers to the science of growing vegetables and fruits such as grapes, apples, etc.;[92] and aquaculture, which relates to the science and practice of fish and shellfish farming. It is important to note that conventionally, it was unusual to afford consideration to the accessibility and mobility of persons with disabilities although all the above activities have been allocated spaces or places on farming sites in different parts of the world. However, it would be of informative to investigate whether specific efforts are being made to preserve spaces or places for persons with disabilities so they can participate in each of the above activities. Ascertaining the presence of disability-inclusive measures for those respective farming activities might be relevant for establishing and demonstrating evidence of good practice that policymakers might use to increase the accessibility of similar spaces.

6.7 Floriculture and evidence of good practice for accessibility students with disabilities

The University of Connecticut Center for Geographic Information and Analysis and the University of Connecticut Storrs Campus Center for Students with Disabilities (CSD) instituted a joint map project in 2007 that was aimed at enhancing the accessibility of farm activities to students with disabilities.[93] This initiative was founded on the idea of promoting equal educational opportunity and full participation for persons with disabilities without discrimination.[94] This undertaking hinged on the policy that no qualified person should be excluded from participating in an agricultural university's programs or activities. In that regard, the CSD devised a disability accessibility map aimed at providing services to all students with permanent or temporary disabilities. The result of such a measure is that in future there is a likelihood of guaranteeing a comprehensively accessible university experience in which agriculture students with disabilities have the same access to programs, opportunities, and activities as other individuals that interact with farming-related environments.[95]

The project undertaken by the University of Connecticut indicates the significance of considering an environment that is supportive of accessibility for persons with disabilities when designing landscapes and spaces or places from where the learning, teaching, and working in areas of floriculture are situated.[96] In many parts of the world, there is a lack of laws or regulatory frameworks for guaranteeing disability-inclusive planning by structural engineers and interior designers.[97] In the context of the USA, it is arguable that the implementation of the Americans with Disabilities Act of 1991 (ADA) might have influenced the extent to which there are greater considerations for individuals with disabilities in the planning and delivering of agricultural education services at the University of Connecticut.[98] ADA, which became law in 1990, prohibits discrimination on the basis of disability status and allows people with disabilities to enjoy equal employment opportunities and equal access to state and local government services.[99] The Connecticut map demonstrates that the need for accessible walkways can clearly be seen as features of inclusiveness similar to the approaches that are elucidated by protagonists such as Smith.[100]

However, it might be correct to advance a polemic assertion that such inclusiveness in agricultural education activities is worth emulating by other sectors of the industry. In that regard, it should be appreciated that UN organizations like the FAO must advise other parts of the world to take similar measures in education and vocational training where these accessibility issues are repeatedly overlooked in agricultural program considerations.[101] Unless actions in relation to what Goldsmith has termed "the architectural model of disability,"[102] or some of the eight measures suggested by Smith, are undertaken in constructing floriculture units,[103] there remains the possibility that persons with disabilities will continue struggling to fit in or adapt to spaces and places that are situated within facilities for working and demonstrating educational activities in floriculture.[104] Literature regarding accessible gardening has been used in this section. The reliance on such disability literature is attributable to the nature of the purpose of floral products, which are produced by sectors of floriculture, bearing in mind florist are specialists for designing gardens. It is almost certain that Smith advances the same view as that of Goldsmith, that improving interactions between some activities of agriculture and disability accessibility would be a noble role in ensuring that flowered areas must be accessible to person with disabilities.

6.7.1 Horticulture and evidence of good practice disability accessibility

From onngoing studies demonstrating that horticulture can play a fundamental role in the medical practice of "therapeutic horticulture,"[105] it is clear that fields and gardens should be inclusive of persons with disabilities. Gardening can provide a motivating impetus that enriches the physical, mental, and social aspects of an individual's life.[106] In the context of horticulture, several studies that have been undertaken by the American Horticultural Therapy Association (AHTA) demonstrate that better accessibility to horticulture can have a positive influence on disability.[107] Thus, for medical reasons, there has been an increase in developed states in promoting inclusive accessibility of spaces or places that are designated for horticulture and gardening activities.[108] Medical and therapeutic arguments for promoting disability accessibility might limit such inclusivity to horticulture linked to care homes and it is unlikely to be legally binding or to become nationally and internationally standardized.[109] In the North American context, universally expected architectural considerations regarding disability inclusiveness through the ADA are a classic example of those legally binding compelling norms. Another significant aspect of disability inclusiveness is the context of the interactions between horticulture and persons with disabilities. This inclusiveness lies in the ability to use modern technology to enhance modifications with respect to the needs of individuals with disabilities. To that end, the coordinator of therapeutic horticulture programs at the Zimmerman Sensory Garden commends adapting currently used farming equipment to match the individual needs of persons with disabilities.[110] Studies have demonstrated that this can be achieved by means of duct tape or any other adhesive.[111] In illustrating that point further, it has been suggested that attaching bamboo poles to shovels with the aim of extending the ease of accessing farming equipment is extraordinarily effective.[112] Someone in a wheelchair can thus be in possession of a ground planter and is in a much better position to comfortably take part in gardening activities.[113] Places where there are steps must be made more disability friendly.

In the above context, most of the suggestions made by Smith could be equally useful in enhancing the inclusivity of agricultural places for people with disabilities.[114] Among these considerations would be increasing walkway spaces to make them more accessible to persons with disabilities (see Figure 6.1). Smith also notes that plant beds ought to be placed at considerably more accessible heights, as demonstrated in Figure 6.2.

Figure 6.1 Walkways designed for inclusiveness for all visitors to enjoy that enhance disability inclusiveness. Tara Turkington (flowcomm). Copyright approval secured on May 12, 2018, https://www.flickr.com/search/?text=Tara%20Turkington%20.

Figure 6.2 Plant beds raised to afford accessibility for agricultural workers. Tara Turkington (flowcomm). Copyright secured on May 12, 2018, https://www.flickr.com/search/?text=Tara%20Turkington%20.

Figure 6.3 Tools and equipment relevant in farming. Tara Turkington (flowcomm).

For those who can bend or sit, there a reasonable likelihood of hardship if this is undertaken for extended periods of time.[115] To that Smith suggests that it would be good for kneelers to be placed next to the ground-level beds to help bridge the gap by offering a resting spot. Consider installing a railing next to the kneelers to help people get up when they're finished.[116] Furthermore, in terms of enhancing the ability to be mobile and to undertake various activities, such rails could play a vital role. With the help of rails, challenges would become much easier to overcome by persons with disabilities.[117]

In the previously mentioned regard, Smith illustrates the importance of disability-adaptive gardening tools (see Figure 6.3).[118] Life with Ease is one of the companies that is known for specializing in supplying disability-adaptive farm equipment.[119] This farm equipment is recommended as their size is designed to improve handling and their textured grips for increasing dexterity in older persons and people with neurological impairments.[120] Special equipment is also available for disabled people with limited hand strength.[121] In the same ethos, adaptive cultivators and shovels allow people to undertake agricultural work if they have a limited range of motion.

6.7.2 *Drone technology and employability of persons with disabilities*

By considering persons with disabilities in the design and development of modern agricultural drones, there would be a better possibility for them to participate in a range of new food production activities.[122] Some of the agricultural activities that could be undertaken by a person with disabilities include: filming real time footage, fence line and bore monitoring, stock locating, asset inspection, checking paddock health, checking irrigation effectiveness, crop monitoring, feed management, surveying for green energy solutions.[123] It is apparent that by using technology on agricultural drones, especially in the context of developed regions, there might be a higher probability for persons with disabilities to be seen as equally participatory to sectors that feed into concepts of agrobiodiversity and agroecology.

6.8 *Recommendations*

6.8.1 *Sustainable development goals as assets in agroecology and agrobiodiversity*

Sustainability requires methods that can consistently empower the skills and dexterity of persons with disabilities. This would mean a greater level of education and greater public awareness of the importance of disability issues in the agricultural context. In the process it becomes apparent that agroecology and agrobiodiversity have a role to play in materializing the benefits of SDGs among persons with disabilities.

6.8.2 *Accessibility of information exchange platforms such as media*

There is a need to ensure that persons with disabilities have better access to ongoing debates. This could include media and journalistic forums at national and international levels that explore matters of agroecology and biodiversity.

6.8.3 *UN agroecology and agrobiodiversity agendas presented in formats that are accessible*

Braille writing as well as sign language have previously been used in international discussions when manifesting inclusive communication on different matters in the UN. However, it is clear that persons with disabilities are customarily distanced from aspects related to activities of the FAO. That would imply that affirmative measures and special considerations might need to be taken into account as a means of spearheading disability activeness rather than passiveness by the different agriculture-related departments of the United Nations.

6.8.4 *Data collection on food security concerning persons with disabilities*

There is currently little information given by states regarding the number of persons with disabilities in regions that have experienced issues of food security and food safety. That information is lacking yet it remains important for purposes of ensuring that clarity and protection are afforded to persons with disabilities.

6.8.5 *Disability consultative surveys on agroecology and agrobiodiversity*

There is a need for surveys on aspects of agroecology and agrobiodiversity that are designed in a manner that is inclusive to persons with disabilities. This inclusiveness will enhance the participation of persons with disabilities in the problem-solving strategies on emerging concerns on agroecology and agrobiodiversity

6.8.6 *Policy evaluation strategies on inclusiveness of laws related to food labelling*

Non-discriminative communication strategies should be found to insure that information and awareness can be conveyed to persons with physical and sensory impairments in relation to food-labeling information concerning organic and inorganic substances.[124] Of course, portraying persons with disabilities as potential customers with purchasing power and client value for food products becomes essential in that context. However, such a perception of persons with disabilities might imply a need to reconstruct discourses relying on charity models of disabilities.[125] In particular, the tendency of using disability charities to concentrate on donating free food to persons with disabilities is problematic, consequently diminishing their perception as individuals that are part of consumer space.[126] Therefore, enhancing food purchasing power will strengthen consumer choices for agricultural products[127] among persons with disabilities while promoting their entitlement to asserting their right to ensure that communication is done through accessible food labelling techniques as far as the world of consumer rights for agricultural products is concerned.

6.9 *Editors' note*

In his chapter, Ivan K. Mugabi focuses on a very specific issue: including persons with disabilities in agroecology- and agrobiodiversity-oriented farming processes. His analysis considers UN literature on this issue, with special reference to the environmental rights of persons with disabilities. He outlines many examples of good practice which apply the principle of inclusive participation as a key tool to develop new farming and agricultural techniques.

In fact, the United Nation's Sustainable Development Goals (SDGs) address disabilities in several of the goals. Specifically,

> in various parts of the SDGs targets, including in relation to: education (to ensure equal access to all levels of education, including for persons with disabilities); growth and employment (to promote decent work and equal pay for all); inequality (as Goal 10 calls for the empowerment and inclusion of all regardless of any status including disability), accessibility of human settlements, as well as data collection and monitoring of the SDGs.[128]

Especially in the agricultural realms, people with disability are at risk of falling into poverty. The UN states that:

> The link between poverty and disability is particularly striking as 80% of persons with disabilities live in developing countries. This also marks disability as a significant cross-cutting issue of the new development agenda. The link between poverty and disability is particularly striking as 80% of persons with disabilities live in developing countries. This also marks disability as a significant cross-cutting issue of the new development agenda.[129]

Implementing practices that welcome and accommodate people with disabilities is a growing concern, especially where "there is a growing recognition within the international community that invisible disabilities, such as mental, psychosocial or development disabilities, are some of the most neglected yet essential conditions to take into account in order to achieve internationally agreed development goals."[130] A lot more work needs to be done to achieve equality and positive upregulation of agroecology and agrobiodiversity may create new and positive opportunities to improve the possibilities for people with disabilities.

Notes

1. Saverio Di Benedetto, "Agriculture and the environment in international law in law: toward a new legal paradigm," in *Law and Agroecology: A Transdisciplinary Dialogue*, Massimo Monteduro, Pierangelo Buongiorno, and Saverio Di Benedetto (eds) (Berlin: Springer, 2015), pp. 99–126.

2. Convention on the Rights of Persons with Disabilities (CRPD) 2515 U.N.T.S. Article 3 on Principles underpinning the Convention.
3. Alex Ghenis, "Disability and climate change: preparing for the future," original posts from the New Earth Disability blog, September 15, 2014. https://worldinstituteondisabilityblog.files.wordpress.com/2016/12/ned-blog-disability-preparing.pdf.
4. "The impact of climate change on people with disabilities," Global Partnership for Disability and Development (GPDD) and World Bank, Human Development Network: Social Protection/Disability and Development Team, July 8, 2009, p. 4.
5. L. Petetin, "Food democracy in food systems," in *Encyclopaedia of Food and Agricultural Ethics*, P.B. Thompson and D. Kaplan (eds.) (Berlin: Springer, 2016), pp. 1–7.
6. Arlene S. Kanter, *The Development of Disability Rights under International Law: From Charity to Human Rights* (London: Routledge, 2015), p. 12. See also Arlene. S. Kanter, "The globalisation of disability right law," *Syracuse Journal of International Law and Commerce* 30 (2003): 241, 269. See also R. Traustadottir, "Disability studies, the social model and legal developments," in *The UN Convention on the Rights of Persons with Disabilities European and Scandinavian Perspectives*, O.M. Arnadottir and G. Quinn (eds) (Leiden: Martinus Nijhoff Publishers, 2009).
7. Alex McClimens and Peter Allmark, "A problem with inclusion in learning disability research," *Nursing Ethics* 18(5) (2011): 633–639.
8. M. Pierri, "Agrobiodiversity, intellectual property right and the right to food," Monteduro et al., *Law and Agroecology*, pp. 451–470. See also John. M. Staatz, Duncan H. Boughton, and Cynthia Donovon, "Food security in developing countries," in *Critical Food Issues: Problems and State-of-the-Art Solutions Worldwide*, Laurel E. Phoenix (ed.) (Santa Barbara, CA: ABC Clio, 2009), pp. 157–176.
9. Olivier De Schutter, "The specter of productivism and food democracy," *Wisconsin Law Review* 2 (2014): 199–234.
10. Paul R. Ehrlich, *The Population Bomb 13–37*, 2nd ed. (New York: Ballantine Books, 1969).
11. "The impact of climate change on people with disabilities," p. 4. See also Jim Bingen, Kathryn Colasanti, Margate Fitzpatrick, and Katherine Nault, "Urban agriculture," Phoenix, *Critical Food Issues*, pp. 109–122.
12. Pierri, "Agrobiodiversity," pp. 451–470.
13. Collin Barnes, "Independent living: a social model account," in *Understanding Health and Social Care: An Introductory Reader*, Julia Johnson and Carinne De Souza (eds), 2nd ed. (Milton Keynes: Open University Press, 2008), p. 88.
14. Tom Penman and Madeline Cruice, "Involving people with communication disabilities in health care institutions," in *Communication in Healthcare*, Karen Bryan (ed.) (Oxford: Peter Lang, 2009), pp. 169–204.
15. CRPD. See also Paul T. Jaeger and Cynthia Ann Bowman, *Understanding Disability: Inclusion, Access, Diversity, and Civil Rights* (Westport, CT: Praegar Publishers, 2005), p. 124.
16. Massimo Monteduro, "Environmental law and agroecology: transdisciplinary approach to public ecosystem services as a new challenge for environmental legal doctrine," *European Energy and Environmental Law Review* 22(1) (2013): 2–11.

17. M. Pierri, "Agrobiodiversity, Intellectual Property Rights and the Right to Food," Monteduro et al., *Law and Agroecology*, pp. 451–470.
18. John M. Staatz, Duncan H. Boughton, and Cynthia Donovon, "Food Security in Developing Countries," Phoenix, *Critical Food Issues*, pp. 157–176.
19. H. Harlan, "Adjucation or Empowerment: Contrasting Experiences with a Social Model of Disability," in *Disability Politics and Struggle for Change*, L. Barton (ed.) (London: Routledge, 2013), pp. 59–78.
20. Emilie K. Aguirre, "Sickeningly sweet: analysis and solutions for the adverse dietary consequences of European Agricultural Law," *Journal of Food Law and Policy* 11 (2) (2015): 252–310. Cf. L. Petetin, "The EU Common Agricultural Policy: towards a more sustainable agriculture?" in *Food and Agricultural Law: Readings on Sustainable Agriculture and the Law in Nigeria*, D. Olawuyi and R. Ako (eds) (Ado Ekiti, Nigeria: Afe Babalola University Press, 2015), pp. 201–224.
21. Aguirre, "Sickeningly sweet." Cf. A. Isoni, "The common agriculture policy: achievements and future prospects," Monteduro et al., *Law and Agroecology*, pp. 185–206.
22. Carlos M. Correa, "Implementation of the TRIPs Agreement in Latin America and the Caribbean," *European Intellectual Property Review* 19(8) (1997): 435–443.
23. CRPD Article 12(5).
24. Anne Saab "Climate-resilient crops and international climate change adaptation law," *Leiden Journal of International Law* 29(2) (2016): 503–528.
25. S. Vernile, "Agriculture, climate change and law," Monteduro et al., *Law and Agroecology*, pp. 421–438.
26. Sylvestre Yamthieu, "The search for a balance between the legitimacy of industrial property rights and the need for food security," *European Intellectual Property Review* 38(9) (2016): 551–562. cf. M.E. Footer, "Intellectual property and agrobiodiversity: towards private ownership of the genetic commons," *Yearbook of International Environmental Law* 10 (1999): 48–81.
27. IPCC, *Climate Change 2007: Synthesis Report. Contribution of Working Groups I, II and III to the Fourth Assessment Report of the Intergovernmental Panel on Climate Change*, [R.K. Pachauri and A. Reisinger (eds) (core writing team)] (Geneva: Intergovernmental Panel on Climate Change, 2007). Cf. "The impact of climate change on people with disabilities," Global Partnership for Disability and Development (GPDD) and The World Bank, Human Development Network: Social Protection/Disability & Development Team, July 8, 2009, p. 4.
28. Noel A. Ysasi, Irmon Marini, and Debra A. Harley, "Climate and weather in the United States and its impacts on people with disabilities in rural communities," in *Disability and Vocational Rehabilitation in Rural Settings: Challenges to Services Delivery*, Debra A. Harley, Noel A. Ysasi et al. (eds), 1st ed. (Berlin: Springer, 2018), pp. 367–382.
29. FCCC/CP/2015/L.9/Rev.1, Conference of the Parties, 21st Session, Paris, November 30 to December 11, 2015. See also FCCC/CP/2010/7/Add.1, Report of the Conference of the Parties on its 16th Session, held in Cancun from November 29 to December 10, 2010. Cf. S. Vernile, "Agriculture, Climate Change and Law," Monteduro et al., *Law and Agroecology*, p. 425.
30. M. Pierri, "Agrobiodiversity, intellectual property rights and the right to food," Monteduro et al., *Law and Agroecology*, pp. 451–470.

31. Ibid., pp. 451–470.
32. Kanter, *The Development of Disability Rights under International Law*, pp. 26–28. Cf. Ida Elizabeth Koch, "From invisibility to indivisibility: The International Convention on Rights of Persons with Disabilities," *in The UN Convention on the Rights of Persons with Disabilities: European and Scandinavian Perspectives*, pp. 67–76.
33. Pierri, "Agrobiodiversity, Intellectual Property Rights and the Right to Food," pp. 451–470.
34. Tesh W. Dagne, "Beyond economic considerations: (re)conceptualising geographical indications for protecting traditional agricultural products," *International Review of Intellectual Property and Competition Law* 46(6) (2015): 682–706.
35. Juliet C. Rothman, "The challenge of disability and access: reconceptualising the role of the medical model," in *Controversies and Disputes in Disability and Rehabilitation*, Roland Meinert and Francis Yuen (eds.) (London: Routledge, 2012), pp. 126–154.
36. Michael L. Perlin, *International Human Rights and Mental Disability Law: When the Silenced Are Heard* (Oxford: Oxford University Press, 2012), p. 33.
37. Pawarit Lertdhamtewe, "Intellectual property law of plant varieties in Thailand: a contextual analysis," *International Review of Intellectual Property and Competition Law* 46(4) (2015): 386–409.
38. P. Buongiorno, "Agriculture, environment and law between ancient experience and present knowledge: some remarks," Monteduro et al., *Law and Agroecology*, pp. 87–99. See also Di Benedetto, "Agriculture and the environment in international law," ibid., pp. 99–126.
39. S. Vernile, "Agriculture, Climate Change and Law," Monteduro et al., *Law and Agroecology*, pp. 421–438.
40. Barbara M. Altman and Sharon N. Barnartt, *Environmental Contexts and Disability* (Bingley: Emerald, 2014).
41. UN committee on the Rights of Persons with Disabilities, "The forgotten victims of Syria's conflict," www.ohchr.org/EN/NewsEvents/Pages/DisplayNews.aspx?NewsID=13736&LangID=E, accessed October 30, 2017.
42. UN Human Rights Council, "Thematic study on the rights of persons with disabilities under Article 11 of the Convention on the Rights of Persons with Disabilities, on situations of risk and humanitarian emergencies," November 30, 2015, A/HRC/31/30, www.refworld.org/docid/56c42c744.html, accessed October 10, 2017.
43. "Myanmar: Cyclone Nargis 2008, facts and figures," www.ifrc.org/en/news-and-media/news stories/asiapacific/myanmar/myanmar-cyclone-nargis-2008-facts-and-figures/, accessed January 5, 2017.
44. Food and Agriculture Organization of the United Nations (hereafter FAO), *World Food Summit* (Rome: Food and Agriculture Organization, November 13–17, 1996), available at www.fao.org/wfs/index_en.htm, accessed October 17, 2017.
45. Ibid.
46. CRPD, Article 5.
47. CRPD, Preamble, para. K.
48. CRPD, Preamble, para. Y. See also Elizabeth Mcgibbon, "Oppressions and access to health care: deepening the conversation," in *Social Determinants of Health: Canadian Perspectives*, Dennis Raphael (ed.), 3rd ed. (Toronto: Canadian Scholars Press Inc., 2016), p. 496.

49. Saksham Chaturvedi and Chanchal Agrawal, "Analysis of farmer rights: in the light of Protection of Plant Varieties and Farmers' Rights Act of India," *European Intellectual Property Review* 33(11) (2011): 708–714.
50. FAO, *Working in Support of Persons with Disabilities*, prepared for the Eighth Session of the Ad Hoc Committee on a Comprehensive and Integral International Convention on Protection and Promotion of the Rights and Dignity of Persons with Disabilities, New York, August 14–25, 2006.
51. Wim Polman and Johanne Hanko, *A Handbook for Training of Disabled on Rural Enterprise Development* (Bangkok: Food and Agriculture Organization of the United Nations Regional Office for Asia and the Pacific, 2003).
52. Global Forum on Food Security and Nutrition, Contributions to Discussion No. 58, "Promoting inclusion of people with disabilities in food security and agricultural development programmes and policies," James Edge, Communications Officer, Food and Agriculture Organization of the United Nations, Rome, pp. 2–21.
53. Carmen G. Gonzalez, "Seasons of resistance: sustainable agriculture and food security in Cuba," *Tulane Environmental Law Journal* 16 (2003): 685–732.
54. Ibid.
55. Ibid., pp. 2, 7, 12, 13.
56. Ibid., pp. 5, 17
57. Ibid., p. 7
58. Ibid., pp. 11, 13.
59. Ibid., pp. 6, 8.
60. Ibid., p. 8.
61. Ibid., p. 18
62. Ibid., p. 19.
63. Ibid.
64. Ibid., p. 10.
65. Ibid., p. 3.
66. Ibid.
67. FAO, *Working in Support of Persons with Disabilities.*
68. CRPD, Article 1, para. 2.
69. FAO, *Putting Ability before Disability in Thailand and Cambodia*, available at www.fao.org/english/newsroom/highlights/2000/001106-e.htm, accessed October 17, 2017.
70. Polman and Hanko, *Handbook for Training of Disabled on Rural Enterprise Development*, pp. 33–34.
71. Glacia Ethridge, David Staten, Kayla D. Goodman, and Delia R. Kpenosen, "Agriculture, farm and immigration workers with disabilities," in Harley et al., *Disability and Vocational Rehabilitation in Rural Settings*, pp. 269–280.
72. Polman and Hanko, *Handbook for Training of Disabled on Rural Enterprise Development*, pp. 33–34.
73. See 48/96, Standard Rules on the Equalization of Opportunities for Persons with Disabilities, March 4, 1994.
74. C. Branes, M. Oliver, and L. Barton, *Disability Studies Today* (Cambridge: Polity Press, 2002), p. 15.
75. United Nations Convention on Rights Persons with Disabilities, Article 1, para. 2.
76. Hans Morten Haugen book reviews at *European Journal of International Law* 23(4) (2012): 1175–1183.

77. Polman and Hanko, *Handbook for Training of Disabled on Rural Enterprise Development*, pp. 33–34.
78. Johanne Hanko, *Mushroom Cultivation for People with Disabilities: A Training Manual* (Bangkok: Food and Agriculture Organization of the United Nation Regional Office for Asia and the Pacific, Thailand, 2001), p. 12.
79. Ibid.
80. T. Degener, "A New Human Right Model of Disability," in *The United Nations Convention on the Rights of Persons with Disabilities: A Commentary*, V. D. Fina et al. (eds) (Berlin: Springer, 2017), pp. 42–57.
81. Muriel Lightbourne, *Food Security, Biological Diversity and Intellectual Property Rights* (New York: Routledge, 2016), chap. 4 ("Equity"), p. 121. Cf. Mugabi Ivan, *Global Health and Public International Law: The Efficacy of Human Rights Treaties in Addressing Health Inequalities in Sub-Saharan Africa, The Case Study of Uganda* (Saarbrücken: Lambert Academic Publishing, 2016), pp. 12–29.
82. Jean Garner Stead and W. Edward Stead, *Sustainable Strategic Management*, 2nd ed. (London: Routledge, 2017), pp. 49–51.
83. C. Ford Runge, Benjamin Senauer, Philip G. Pardey, and Mark W. Rosegrant, *Ending Hunger in Our Lifetime: Food Security and Globalization* (Baltimore, MD: Johns Hopkins University Press, 2003), pp. 205–206.
84. Great Britain Parliament, Hansard, House of Commons, written answers to questions, January 20, 2004, col 1184W, https://hansard.parliament.uk/Commons/2004-01-20/debates/297bfe17-a304-4743-9da4-d212d043d1fc/WrittenAnswers.
85. Stead and Stead, *Sustainable Strategic Management*, pp. 49–51.
86. Danial W. Wong and Lucy W. Hernandez, "The Asia Pacific region: rural urban impact on disability," in *Disability and Vocational Rehabilitation in Rural Settings: Challenges to Services Delivery*, Debra A. Harley, Noel A. Ysasi et al. (eds), 1st ed. (Berlin: Springer, 2018), pp. 317–334.
87. Ellen K. Cromley, *Accessibility Map, University of Connecticut, Storrs and Depot Campuses* (Storrs, CT: University of Connecticut Centre for Geographic Information and Analysis, 2006).
88. Selwyn Goldsmith, *Designing for the Disabled: The New Paradigm* (London: Routledge, 2011), p. 145.
89. Melissa Bravo, *A Guide for Making Community Gardens Accessible for All Members* (Buffalo, NY: Grassroots Gardens, 2015), p. 18.
90. Hazel White and Matthew Plut, *Paths and Walkways: Simple Projects, Contemporary Designs* (San Francisco, CA: Chronical Books, 1998), p. 18.
91. S.E. Smith, "'8 Ways to Make Gardening Accessible for People with Different Needs'" May 19, 2013.
92. Mark Alan Christie, Michaela Thomson, Paul K. Miller, and Fiona Cole, "Personality disorder and intellectual disability: the impacts of horticultural therapy within a medium-secure unit," *Journal of Therapeutic Horticulture* 26(1) (2016): 3–17.
93. Cromley, *Accessibility Map*.
94. Ibid.
95. Edith Brown Weiss, *Environmental Change and International Law: New Challenges and Dimensions* (Tokyo: United Nations University Press, 1992).
96. Bravo, *Guide for Making Community Gardens Accessible*, p. 18.
97. Goldsmith, *Designing for the Disabled*, p. 145.

98. Bravo, *Guide for Making Community Gardens Accessible*, p. 18.
99. Ibid., pp. 3–4.
100. Smith, "8 Ways."
101. Bravo, *Guide for Making Community Gardens Accessible*, p. 18.
102. Goldsmith, *Designing for the Disabled*, p. 145.
103. Smith, "8 Ways."
104. Ellen Cromely, "Accessibility Map of Connecticut, Storrs and Depot," Campuses Centre for Geographic Information and Analysis, 2006.
105. Bravo, *Guide for Making Community Gardens Accessible*, p. 4.
106. Ibid.
107. Claudia K.Y. Lai, Lily Y. W. Ho, Rick Y. C. Kwan et al., "An exploratory study on the effect of horticultural therapy for adults with intellectual disabilities," in *Grow It, Heal It: Natural and Effective Herbal Remedies from Your Garden or Windowsill*, Christopher Hobbs and Leslie Gardner (New York: Rodale, 2013), p. 230.
108. Christie et al., "Personality disorder."
109. Bravo, *Guide for Making Community Gardens Accessible*, p. 4.
110. Ibid.
111. Ibid.
112. Ibid.
113. Ibid.
114. Smith, "8 Ways."
115. Christie et al., "Personality Disorder."
116. Ibid.
117. Melissa, *Guide for Making Community Gardens Accessible*.
118. Hanko, *Mushroom Cultivation*, p. 12.
119. Polman and Hanko, *Handbook for Training of Disabled on Rural Enterprise Development*, pp. 33–34.
120. Ibid.
121. Bravo, *Guide for Making Community Gardens Accessible*.
122. Chris Anderson, "Agricultural drones," *MIT Technology Review* (April 23, 2014), https://www.technologyreview.com/s/526491/agricultural-drones/. Relatively cheap drones with advanced sensors and imaging capabilities are giving farmers new ways to increase yields and reduce crop damage.
123. Dave Heavyside, "Drones in agriculture: the new frontier," The Institute for Drone Technology, Lardner, Victoria, June 24, 2017.
124. Charles Lawson, "Implementing farmers' rights: finding meaning and purpose for the international treaty on plant genetic resources for food and agriculture commitments," *European Intellectual Property Review* 37(7) (2015): 442–454.
125. Carrie A. Scrufari, "Chickens and cows are not the answer: why charity-based models focused on donating livestock will not solve global hunger," *University of Maryland Law Journal of Race, Religion, Gender and Class* 16(2) (2016): 209–238.
126. Elver Hilal, "The challenges and developments of the right to food in the 21st century: reflections of the United Nations Special Rapporteur on the Right to Food," *UCLA Journal of International Law and Foreign Affairs* 20(1) (2016): 1–44.

127. Kanchana Kariyawasam, "Access to biological resources and benefit-sharing: exploring a regional mechanism to implement the Convention on Biological Diversity (CBD) in SAARC Countries," *European Intellectual Property Review* 29(8) 2007: 325–335.
128. United Nations, *Equal Rights for Persons with Disabilities: Key to Achieve New Development Agenda* (June 14, 2016), https://www.un.org/sustainabledevelopment/blog/2016/06/equal-rights-for-persons-with-disabilities-key-to-achieve-new-development-agenda/.
129. Ibid.
130. Ibid.

chapter seven

The special case of olives

Alexander Cherry

Contents

7.1 Introduction

The olive is one of the most important crops of human history. Consumed both in "table" form as a pickled fruit and, more commonly, pressed into oil, it has been used as a culinary staple, in cosmetics, medicines, rituals, as an embalming agent, and as a source of biofuel and light throughout the Mediterranean and Middle East for thousands of years. Long-lived, hardy, and heavily drought-resistant, olive trees are typically managed in monoculture groves over the course of multiple human generations. In Ancient Greece, olive trees and oil were regarded as sacred. According to myth, the city of Athens was named after Athena because her gift of the olive to the inhabitants was more valuable than Poseidon's offering of the horse.[1] Amphorae full of high-quality oil were given as prizes to contestants in the ancient Olympic games. The ubiquity and cultural importance of the tree are attested in the writings of countless famous playwrights and historians and on artifacts found throughout the region.[2]

It is generally believed by historians and archaeologists that olive trees were introduced to Greece via Crete as early as 3000 BC from the Middle East.[3] From there they were quickly spread by Greeks throughout their colonies around the Mediterranean. The spread of olive oil throughout Europe, the Middle East, and North Africa was completed by the Roman empire, which assigned it a high cultural and economic value. Although some of the uses of olive oil changed over time, it always remained one of the most important crops throughout the ancient world. This tradition was carried on both through the Catholic and Eastern Orthodox churches, but also the Muslim caliphates, even after other sources of fuel and cooking oil were developed. To this day, in Mediterranean countries, olive oil remains one of the most important agricultural products and culinary oils and is still used in cultural events including baptisms and weddings.[4]

7.2 Current olive oil production

Today, Greece is the world's third largest producer of olive oil, after Spain and Italy. Over 60 percent of Greek agricultural land is dedicated to olive groves.[5] The tree grows especially well in the climate (hot, dry summers and cool winters) and rocky soils of the region. Although the practice of growing olives is ancient, the process has become heavily modernized in recent years, with modern pesticides, fertilizers, and irrigation in widespread use. Unlike Spain and Italy, however, where massive monoculture olive groves generate the largest proportion of a national production that is transported around the country or exported after processing, Greece's net production is dominated by small-scale producers for local consumption.[6]

Net olive production fluctuates yearly based on local climatic factors as well as the trees' natural yearly oscillations in yield. Dry weather is ideal so as to not hurt the trees during the sticking process. If the weather is wet and rainy, fungal diseases may develop as well. Too much rain in the spring may destroy the flowers and reduce the number of fruits developed for the entire year. Harvesting occurs once per year. The farmers can decide when to harvest the fruits based on the microclimate, which is heavily affected by altitude and precipitation, as well as by how much acidity they would like.[7] The purest, high-quality oil with low acidity is typically yielded by olives harvested in late October, about two weeks after the arrival of autumn rain, so that the moisture has time to be absorbed by the tree's roots and carried up to the fruits.[8]

In the winter and spring months, the farmers prune the trees, clear grasses growing below them using heavy machine equipment, spray pesticides, add fertilizer to the soil, and introduce water via irrigation or rainwater collection systems.[9] In addition to fungal infections, the olive fruit fly is the most destructive and widespread pest affecting olive farms in the Mediterranean. Adult flies lay eggs inside the fruits, which the larvae then eat as they grow, causing premature dropping and often destroying up to 50 percent of the crop.[10]

The processing of olives into oil has been heavily mechanized in recent years. After careful harvesting and transporting, twigs and stems are removed, and the olives are washed. Then, the olives and pits are ground together into a paste and slowly mixed with water. Oxidation is to be avoided during this process but may occur if the mixture is heated, which speeds up the process. "Cold-pressed" or "virgin" oils are therefore less oxidized, with lower acidity, and considered to be of higher quality but can be only produced in lower quantities. The paste is then sent through a centrifuge to separate the liquid from solid matter, and the oil is separated from the water. Finally, the oil can be refined further to change the acidity, color, or odor, typically for sale in international markets.[11] Some factories may burn the solid waste byproduct as biofuel to run their machinery.[12] Generally, the waste products including waste water are left to decompose in highly concentrated lagoons, which constitute a major source of environmental pollution in the regions where they are produced.[13]

Although these modernization and mechanization changes in the ancient olive industry (as well as increased land use) have increased the net production of olive oil, there have been some unintended negative consequences as well. Chemical fertilizers and pesticides leach harmful chemicals into the groundwater and runoff water systems that end up damaging the drinking water supply as well as surrounding natural ecosystems. Furthermore, chemicals sprayed on the olives themselves to protect against pests may be harmful to human health. In response, there has been a growing movement of so-called "alternative" production methods such as organic or biological farming, which are really just a return to the older indigenous practices, and increased demand for products made with these methods.[14]

7.2.1 The case of Samos Island

Samos is a somewhat large, green, mountainous, and biodiverse Greek island in the Eastern Aegean Sea, home to 33,000 people,[15] although the number swells greatly during the summer months. It is most famous for being the birthplace of the Ancient Greek philosopher and mathematician Pythagoras, the birthplace of the goddess Hera, according to Greek mythology, and the Greek island closest to mainland Turkey, separated by the 1.7 kilometer-wide Mycale Strait. The local economy primarily revolves around tourism and agriculture, and the 115 square kilometers of olive groves cover 25 percent of the total land area of the island.[16]

As in the rest of Greece, and especially other Aegean islands, olive production is dominated by small-scale producers and families who own a small grove of a few trees and produce for their own consumption throughout the year, although many families also have larger groves as well. The average grove is 10 acres. It is rare for any producer to produce only olives, though the groves are typically separate from other crops. Irrigation is rare, though chemical fertilizers and pesticides are frequently used.[17] In fact, the municipality governing the island sprays the entire island with pesticides to combat the olive fly.[18]

The island is home to a growing subset of olive growers using organic methods, motivated by both their personal beliefs in the older, more "natural" methods as well as to supply the demands of wealthy tourists from Northern Europe who visit the island in large numbers throughout the summer. Organic farmers are limited by the types of pesticides and fertilizers that they are allowed to use, and their yields per acre tend to be lower than non-organic producers.[19]

There are over ten small factories on the island that take olives harvested from commercial as well as small family growers and process them into oil. For the small families, the factory typically takes somewhere between 10 percent and 15 percent of the resultant oil as payment and returns the rest to the family. A new tax policy is in the process of being implemented, however, that will require growers whose primary source of income is not farming to pay a higher tax percentage of their product. This policy will likely lead many of these families with other sources of income to cease to harvest and bring their olives to the factory, and thus exit the rural economy.[20]

As with much of the rest of the economy on Greek islands, the network of growers and processors is primarily maintained through personal and familial relationships. During the several hours-long crushing and pressing processes, the clients will pass the time together chatting and playing cards or backgammon, drinking coffee or ouzo. This social gathering is a chance as well for the various producers in the community to exchange information about the productivity of their agricultural efforts throughout the year, among other things.[21]

In recent years, the international Agrocert certification body has approved the creation of a Protected Geographical Indication covering olive oil produced on the island and the first factory has been certified organic by the Greek authorities.[22] The unique microclimate of Samos results in a high-quality, aromatic oil, using olives native to the island. When cooked at a temperature below 27 degrees Celsius (cold-pressed), the nutrition and taste quality of the oil is best preserved, and the "extra-virgin" label can fetch much higher prices in the market.[23]

After returning the proper amounts of olive oil to the small family producers who brought their supply to the factory, the rest of the finished product primarily ends up in one of two main destinations. The first is to supermarkets and tourist shops around the island, and the second is to international purchasers. Samian factories have been working in recent years to promote the distinctive qualities of the oil produced on the island in order to carve out a niche in the international market, since nearly all regions of Greece itself produce more olives than they can consume locally.[24] This international focus is quite recent, however, and due to the insularity of the island and dominance of informal networks in business and cultural life, it can be difficult for outsiders to infiltrate the inner workings of the industry, although people are quite friendly and open once a connection is established.

7.2.2 Threats to the olive economy in Samos

The future prospects concerning the health and sustainability of the olive oil industry seem to be in good shape, although there will definitely be some major challenges. Over the last five to ten years the number of refugees illegally crossing from Turkey to Samos in order to claim asylum in the European Union has skyrocketed, and some locals blame them for the decline in the tourists visiting the island every year, although there is some evidence to suggest that this decline was occurring even before the refugees arrived.[25] In any case, there has been a shift in the local economy to make up for the decline in tourism toward more-expensive agricultural products that can be sold internationally, including olive oil. There is still much work that can be done to increase the monetary gains from these efforts even without increasing production, simply by reaching a wider audience and distinguishing the local product from the various other heritage olive producers that can be found everywhere in Greece.[26]

One of the biggest obstacles to these efforts is the aging and declining population, with low fertility rates and most young people leaving to find more economic opportunities in Athens or Thessaloniki, the two largest cities in Greece. If the next generation does not have enough people to sustain the ancient agricultural production, there is a danger that it could die out.

The new policy that would raise the tax that small families must pay to oil factories in order to have their olives processed may cause many families to decide that the effort is no longer worthwhile to harvest their own olives, choosing instead to purchase their oil needs from supermarkets. The purpose of this tax is ostensibly to help the factories, which are trying to raise their profit margins and put Greek oil higher on the international market, but it may negatively impact the local communities where the olives are produced in favor of larger-scale producers.[27]

Climatic changes, including droughts, increased wildfires, increased likelihood of extreme weather events such as frost or snow in winter, and changes in rainfall patterns, are also large threats to olive production on the island as well as around the entire Mediterranean. Since olive trees fluctuate yield yearly and are highly sensitive to small climatic changes, global climate change can have massively negative consequences on the industry. Furthermore, these changes may increase the populations of pests such as the olive fly that will require higher amounts of expensive and environmentally damaging pesticides in order to maintain an economically viable yield of olives per year. A 2014 analysis revealed that small producers on sloping lands in areas prone to aridification in Greece, Italy, and the Middle East are at the highest risk for heavy economic losses due to climate change, and Samos's terrain fits into these categories.[28]

7.3 Conclusion

In short, olive oil, one of the Mediterranean's most important agricultural products, has a long history and massive significance even to this day. In Greece and particularly the island of Samos, olive oil production is dominated by small family producers which are threatened by climatic and economic changes. However, the rise of higher quality and more sustainably produced olive oils bodes well for both the environment and the larger producers and factories that stay in the industry.

7.4 Editors' note

This chapter describes the importance of olive oil production in agricultural systems, and outlines a significant case study, with reference to olive oil production in the Greek island of Samos.

Alexander Cherry points out the great importance of olive oil for the Greek economy as the third producer in the world after Spain and Italy, with the specific characteristic of small producers for local consumption. After describing the production process, with special reference to the best time for the harvest, he notes that in recent years a movement for "alternative" production represented a new trend in olive oil production.

In the case study on Samos Island, he outlines a clear distinction between traditional methods, which involve significant use of pesticides, with specific reference to combat against the olive fly, and the new methods used by some producers, which prefer natural methods and organic farming solutions.

The divide in farming techniques described by Alex Cherry represents, in our view, the future of olive oil production. The olive fly is a major issue in all Mediterranean countries due to climate change, and the agriculture techniques which combat this problem without the use of pesticides must be considered as the best solution to be promoted among producers.

New regulation must introduce, as a general principle, the right of the consumer—as a potential buyer—to access detailed information on the use of pesticides in agricultural products aimed at consumption. New blockchain techniques for agricultural product labelling may represent a model for the regulators, especially in the olive oil industry, where traceability is a requirement easily achieved from a technical standpoint.

Notes

1. Hartmann, Hudson T., and Plato G. Bougas, "Olive Production in Greece," *Economic Botany* 24(4) (1970): 443–59, doi:10.1007/bf02860750.
2. Ibid.
3. Ibid.
4. Ibid.
5. "Olive Oil Production in the Mediterranean., PROSODOL (2012). Accessed November 23, 2018, www.prosodol.gr/?q=node/203.
6. European Union, Agriculture and Rural Development, Directorate-General for Agriculture and Rural Development, *EU Olive Oil Farms Report: Based on FADN Data* (Brussels: European Commission, 2012). Accessed November 23, 2018, https://static.oliveoiltimes.com/library/olive-farms-report.pdf.
7. Nikolaou, Anastasia. Email interview by author (November 2018).
8. Delaveri, Kleopatra. In-person interview by author (October 2018).
9. Nikolaou, interview.
10. Ponti, L., Andrew Paul Gutierrez, Paolo Michele Ruti, and Alessandro Dell'Aquila, "Fine-scale Ecological and Economic Assessment of Climate Change on Olive in the Mediterranean Basin Reveals Winners and Losers," *Proceedings of the National Academy of Sciences* 111(15) (2014): 5598–5603, doi: 10.1073/pnas.1314437111.
11. Alleman, Gayle A., "Ultimate Guide to Olive Oil," HowStuffWorks. December 27, 2006. Accessed November 23, 2018. https://recipes.howstuffworks.com/how-olive-oil-works1.htm.
12. Delaveri, interview.
13. "Olive Oil Mills Wastes," PROSODOL (2012). Accessed November 23, 2018, www.prosodol.gr/?q=node/453.
14. Nikolaou, interview.
15. 2011 Government Estimate.
16. "Olive Cultivation Constitutes the Basic Form of Cultivation on Samos," SamosIn. Accessed November 23, 2018, www.samosin.gr/cultivations-of-samos/olive-cultivation-at-samos/.
17. Nikolaou, interview.
18. Delaveri, interview.
19. Nikolaou, interview.
20. Ibid.
21. Delaveri, interview.
22. "Olive Cultivation Constitutes the Basic Form of Cultivation on Samos."
23. "The Valsamidis Bros and Theorema [Theorem] about Samian Olive Oil," My Samos: Ένα Blog για τη Σάμο. Accessed November 23, 2018, http://my-samos.blogspot.com/2017/07/blog-post_72.html.
24. Nikolaou, interview.
25. Spilanis, Ioannis, H. Vayanni, and K. Glyptou. "Evaluating the Tourism Activity in a Destination: The Case of Samos," SAPIENS (Surveys and Perspectives Integrating Environment and Society) (December 10, 2012), Accessed November 23, 2018, https://journals.openedition.org/etudescaribeennes/6257?lang=en.
26. Ibid.
27. Ibid.
28. Ponti et al. "Fine-scale Ecological and Economic Assessment."

chapter eight

South Asian perspectives of food and law in agroecology and agrobiodiversity

Sumit Saurav

Contents

8.1 Introduction

Global concern on environmental issues and promotion of the "sustainable development" model to sustain ecological diversity and natural resources for future generations has been resolved in the Stockholm Declaration, which was eventually emphasized in the Rio De Janeiro Summit under "Agenda 21." Scientists have long settled the impact on the economy due to depletion of natural resources, diminishing biodiversity, global warming, famine, food security, etc. Hence, the balancing framework between development and ecology becomes the strategized policy across the globe in furtherance of socio-economic development programs. Increasing dimensions of developmental models have escalated manifold complexities in the legal frameworks to define those and set necessary regulations to uphold the dispensation of justice. The Paris Summit has sent a strong message regarding the perceived threat of global warming and climatic changes. Again, the progressive move toward the worldwide convergence on Corporate Social Responsibility has implored the active participation of business in saving the planet as well. Different governmental and non-governmental organizations are promoting different economic incentives for renewable use of natural resources. Tariffs on carbon emission are being prompted with stern action to prevent the most harm to nature. On the other hand, the competition regime has accelerated the competitive ambiance amongst businesses to exploit natural resources to capture maximum market share. Price volatility in the agro-produced-markets from climatic change is causing harm to the livelihood of the many communities across the globe. The universal approach, "sustainable development" across the globe, has to be appreciated from the perspective of the shortcomings of different domestic socio-economic and socio-legal contexts due to polarized economic powers.

8.2 Defining food and its law and usage

Promoting the idea of "sustainable development" from economic prosperity by exploiting natural resources to achieve maximum utility in the global competitive landscape has hidden the consequential impact upon the ecology. This exploitation has resulted in climate change, diminishing biodiversity, global warming, and ecology imbalance, which has, in turn, begun reshaping the social environment, technologies, and behavioral patterns concerning the perceived threat to the natural habitats and species, food security, famine, etc. at a snail-like pace. "Sustainable development" upon implementation has evinced positive socio-economic consequences. However, over time, ineptitude and lack of prioritizing more concrete policies required to sustain a balanced ecology have confounded the economy and environmental conservation. In this chapter, I highlight the research and interpretation of features of the environment that have changed over time and consider future access to natural resources that may endanger not only human society but also the economy. Plainly speaking, it is true that demand–supply forces always reveal the inadequacy of the resources—sometimes due to the unavailability of land and natural resources with the population increasing population and at times due to unexplored dimensions compatible with human utilities.

Nevertheless, since the state is not responsible for curing the central issue, in terms of both economical and federal governance, the interpretation of the court is required to achieve the maximum utility. And if that fails, then it may cause a problem.[1] Rights are the legally protected and recognized interests of the individual and imply a standard of behavior of the individual. It is a universal principle that arguably inspires the thinking of human rights. When one claims a right, there is always a corresponding duty imposed on other; but the opposite appears to be true. The law must determine the limits imposed by these duties, and must be respectful of the rights of others and meet "the just requirements of morality, public order and the general welfare in a democratic society."[2]

During the nineteenth century, the idea of "commoners' rights" was a fallacy. However, the logical and continual culmination of the public voice in the course of time exhorted the state to endeavor to conserve the vulnerability of individuals. The leadership of the governing council in colonial India notably promoted commercial exposure, especially to enhance a vibrant supply chain for the home country. As a result, India faced famine. Food stock reached an unprecedented low for citizens, while hunger in drought-affected areas intensified. Food security for the nation to protect the state from chronic hunger has now been endorsed by the Supreme Court of India after five decades of independence in its judgment: *People's Union for Civil Liberties v. Union of India & Others.*[3] However, the impact of climate change and increased pollution have not been acknowledged sufficiently by the state. The above judgment introduced the "right to food" by a number of interim orders. After the first order, on November 28, 2001, the government introduced the Public Distribution System, Antyodaya Anna Yojana, National Programme of Nutritional Support to Primary Education (midday meal scheme), Integrated Child Development Services and Annapurna as legal entitlements of the nation. That order instituted: responsibility for compliance,[4] accountability to local bodies,[5] access to information,[6] dissemination of Court orders,[7] schemes should be continued,[8] and full utilization of grain quotas.[9] However, the dynamics of the changing climate may affect the undertaking and frustrate the vision of food security unless the state takes some significant steps as a matter of urgency.

Whenever we talk about the right to food, it creates a question in the mind: "What is it?" Is the government under an obligation to provide free food to every individual under its jurisdiction? This is not so. "Right to food" is shorthand for a more complex set of obligations relating to "food security" which involve ensuring access to food and planning for shortages and distribution problems.[10] In India, its citizens can view the "right to food" as an implication of the fundamental "right to life," enshrined in Article 21 of the Constitution of India. The expression "Life" in this Article has been judicially interpreted to mean a life with human dignity and not mere survival or existence.[11] In light of the aforementioned, the state is obliged to provide for all those "minimum requirements" which must be satisfied to enable a person to live with human dignity: education, healthcare, just and humane conditions of work, protection against exploitation, etc. Articles 39(a) and 47 of the Constitution interpret the concept of the right to food. Article 39(a) directs the state to ensure that all citizens have "the right to an adequate means of livelihood." According to Article 47, "the State shall regard the raising of the level of nutrition and the standard of living of its people and the improvement of public health as among its primary duties." These two Articles belong to the "Directive Principles of State Policy," which are not supposed to be enforceable in court (Article 37).

However, it is possible to argue that Articles 39(a) and 47 are enforceable in court as expressions of the fundamental right to life.[12] Hence, the right to be free from hunger enshrined in Article 21 is to be ensured by the fulfillment of the obligation of the state as laid down in Articles 39(a) and 47. The reading of Article 21 together with Articles 39(a) and 47 places the issue of food security in the correct perspective, thus making the right to food a guaranteed Fundamental Right which is enforceable by the constitutional remedy provided under Article 32 of the Constitution.[13]

While the Indian Constitution does not enshrine the right to food expressly, the Supreme Court of India has come up with Kishen Pattnayak & another v. State of Orissa[14] and People's Union For Civil Liberties v. Union of India[15] wherein has recognized the right to food under the right to life stipulated in Article 21 of the Indian Constitution, with reference also to the Directive Principle of State Policy concerning nutrition, contained in Article 47.[16] The Supreme Court held on May 2, 2003 that state failure to implement food schemes and distribution in cases of starvation and risk of hunger, where grain stocks were available, amounted to a violation of the right to life. They further went on to issue some interim measures prompting the state to implement the Famine Code and detailing a few rules to be complied with, especially in connection with vulnerable groups.[17]

The Supreme Court of Bangladesh, interpreting the constitutional clause as enshrining the right to life, decided that the government should remove threats posed by a consignment of powdered milk which exhibited levels of radiation above the acceptable limits. The court stated that the right to life includes the protection of health and normal longevity of an ordinary human being and that the consumption and marketing of food and drink injurious to health can threaten that right.[18] Litigation on the right to food has since preserved a provision in Article 15 of the Constitution deeming this right as a basic necessity.[19]

Sri Lanka has protected the right to food in clause (2) of Article 27 of the Constitution in chapter IV containing directive principles of state policy and fundamental duties.[20] Pakistan has also enshrined the right to food in Clause (d) of Article 38 under the promotion of the social and economic well-being of the people.[21] By April 2011, the Supreme Court of Nepal published a critical decision clarified by an interim order issued in September 2008 for the immediate provision of the right to adequate food. Aside from the availability of food, the Court emphasizes the role of the authorities in ensuring that food is accessible and affordable for the people.[22]

Almost every court in the South Asia region understands the concept of the right to food and has instituted a relevant law. However, the legislature has failed to appreciate the definition of "adequate" food. Almost every country's legislature has enshrined the right to food in their

Constitution but the Constitution has defined neither the meaning of adequate food nor who are the neediest to claim such right and under what circumstances. The word "adequate" merely means "enough" or "having the requisite qualities or resources to meet a task," which does not clarify what is the requisite quality of the food in question.

8.3 Particular challenges regarding the right to food

Since many scientists and leading experts consider the right to food to mean "food security," it is essential to highlight the issues regarding the concept of "rights." As a result of this "silent tsunami," international organizations and individual countries have embarked on an aggressive food security crusade as the only option to make food affordable to all. An Emergency Food Security Assessment (EFSA) in 2005 determined that over 750 million people were food insecure in 70 poorer countries.[23] Asia and the Commonwealth of Independent States have since experienced a 30 percent drop in the number of hungry people. However, analysts hold the view that high food prices will cause an increase in food insecurity and a widespread food crisis in many developing countries. Poor people in developing countries spend between 50 percent and 80 percent of their incomes on food and this is most relevant with poor rural households.[24] Any increase in the price of food reduces food consumption and increases hunger. The following section is concerned with discussing the issues with regards to the right to food.

The first issue is related to the rights of the individual. Human rights are many and complex. The question here is: "Is there prioritization among different human rights?" In a democratic country, there should be no prioritization regarding different rights. Different types of rights have the effect of mutually reinforcing themselves: better nutrition, health, and education will lead to improvements in political freedoms and the rule of law; similarly, freedom of expression and association can ensure that the best decisions are taken to protect[25] the right to food, health, and work.[26] Therefore, the right to food needs to be properly secured before it turns into a luxury similar to the right to vote or a privilege like the freedom of expression.

The second issue is related to food security planning for the needs of the population. The fundamental purpose of implementing the right to food is to provide food freely to the needy. Here, the neediest are the students and adults who are the future of tomorrow. Nobody can avoid the fact that the South Asia region is far behind other developed or developing countries in literacy. The reason behind this is that the region's children cannot concentrate on their education because they are not getting a proper diet. It is a matter of grave concern as it will attack the economy in the future. So, there is a need to provide free food to every student without any discrimination.

The third issue is related to the quality of the food and again it is a matter of great concern the eating free food is often not enough: a person needs a proper diet. In particular, there should be no violation of the right to food through the unjustified destruction of crops or evictions from land. Moreover, it is the government's duty to provide a proper diet for the victims of any natural calamity, such as when the tsunami hit the State of Tamil Nadu in South India. The government provided food to the people affected but it was not a proper diet. In this particular case, there must be no discrimination about access to food.

The fourth issue is related to public distributive systems (PDSs) which can be criminally subverted. PDS is a means of distributing food grain and other essential commodities at subsidized prices through "fair price shops." But PDS has been compromised, and it is being replaced by direct cash transfer (DCT). Whether DCT will successfully curb corruption remains to be seen. In the present social scenario, DCT is beneficial for urban people but not for the rural population. The success of DCT depends principally on accessibility to banks, and in the case in rural India, the banking infrastructure is extremely limited: people have to travel long distances to access a bank, which may not guarantee food security.

Moreover, the banking system in rural areas is not ready to handle large volumes of small transfers. Banks are often a long way away and overcrowded. The proposed solution—banking correspondents—is fraught with problems. Governments could potentially convert Post Offices into payment agencies, but this will take time. Rural markets are often poorly developed.[27] Even people in rural regions need entitlement, not cash.[28] Dismantling PDS will disrupt the flow of food across the country and put many people at the mercy of local traders and intermediaries. There are particular concerns for special groups such as single women, disabled persons, and the elderly. These groups cannot travel far to withdraw their cash and buy food from distant markets. What is more, inflation could without any trouble erode the purchasing power of cash transfers. When the government refuses to index pensions or National Rural Employment Guarantee Act (NREGA) wages, how can it be trusted to index cash transfers to price levels? Even if some indexation does happen, small delays or gaps in price information could cause significant hardship for poor people.[29]Alas, myths cannot be buried so quickly, and it is comfortable to believe that it is "those lumpens" who inhabit the resettlement colonies and slum areas who are responsible for corruption.

The fifth issue is related to the prices of commodities such as rice, corn, and wheat, which have all reached record highs recently, and pose a significant threat to developing countries. Some factors attributed to the price spike include climate change, population growth, increased demand for bio-fuels, failure to improve crop yield, and high oil and input prices. Speculation on the commodity market, and structural problems like underinvestment in agriculture and infrastructure, the dominance in the supply chain, and food and agricultural policies could also push up prices. As a consequence, food stocks have become depleted. In 2005, food production was dramatically affected by extreme weather incidents in the major food-producing countries. By 2006, world cereal production had fallen by 2.1 percent. In 2007, rapid increases in oil prices increased fertilizer and other food production costs.[30]

The final issue here is related to the shortage of food, resulting in famine. A famine occurs when a large number of people within a community suffer from entitlement failure. It is such a scenario where demand is high but the supply is low because the land is constant but the population is increasing. Under these circumstances, we need to change the economic concept not simply seek the assistance of the courts. An individual is said to suffer from the failure of food entitlement when her entitlement set does not contain enough food to enable her to avoid starvation in the absence of non-entitlement transfers; and in such situations, a famine occurs. An entitlement set is first derived by applying E-mapping on the endowment set. A single change in entitlement set occurs only through differences in either an endowment or E-mapping set which creates entitlement failure and thus famine.[31] A fisherman, for instance, loses his boat, which prevents him from catching fish that he must exchange to get his staple food, and the fish to be exchanged for the minimum amount of rice he needs creates exchange failure. Such a situation occurs when any set disturbs and ultimately leads to trade entitlement failure. Therefore, the point is that we need to eco-balance the three sets of trade or else this will threaten access to food.

8.4 Corporate Social Responsibility and the right to a healthy life

Corporate Social Responsibility (CSR) is a form of insurance undertaken by corporate entities to compensate for losses or damage for which their industries are responsible. It varies according to the damage caused.[32] It may be a supportive hand to maintain the ecological balance of the community, as big businesses are first to disrupt the environment. In other words, it is a type of social insurance where big corporations pool contributions from their profits, depending upon their income or turnover, which is utilized, for example, to rectify the loss of the ecosystem and to ensure a minimum standard of the environment is maintained. It implies social protection which stands aside to men.[33] The concept of CSR is embedded in section 135 of the Companies Act 2013, read with Rule 3 of Corporate Social Responsibility Policy 2014.[34] The legislature has introduced it with the intent of reconstructing losses caused by the subject entity. The provision demands that businesses should spend a minimum of two percent of their average net profits in every financial year in the pursuance of CSR. It may revive the debate of the colonial period whether or not the welfare mechanism of the state could be shifted through the delegation of economic power to private stakeholders in the developmental context. Or whether a company facing a loss should also take part in CSR. However, the USA and other international experts have concluded that the world should invest in minimizing the amount of climate change that occurs rather than adapting to the changes that are avoidable. Moreover, there are a series of judgments wherein extended producer responsibility (EPR) is accepted (which is also known as "the polluter pays principle." What this principle denotes is that a polluter is liable to reconstruct losses. The aim behind the implementation of such a policy is to enhance industrialization and economic development. Further, research shows that increase in demand results in the transformation of the environment and contributes to the destruction of the ecosystem on an extraordinary scale as evidenced by climate change.[35] Hence, the same fund can be utilized for more suitable causes such as restoring devastating environmental issues and detecting a set of policies, practices, or standards of behavior so as to provide long-term economic opportunities and improved quality of life around the world, while maintaining a sustainable environment or viable ecosystem.

Heeding environmental and climate concerns means markets can develop while the environment can be protected through employing cheap insurance or other alternative risk transfer approaches. The judiciary has played their part in helping reconstruct the environment and in protecting the environment from deterioration. In *Ratlam v. Vardhichand*,[36] the Supreme Court of India directed the municipality and owners of alcohol plants to provide a drainage system within one year and to improve other sanitary conditions. In *Rural Litigation and Entitlement Kendra, Dehradun v. State of U.P.*,[37] the Supreme Court for the first time attempted to look into the question of ecological balance. The petitioners of a voluntary organization pleaded that mining activities were the cause of environmental disturbance to the Mussoorie Hills (the mining company's removal of trees and forest cover was resulting in soil erosion which caused blockages to the underground water channels that fed rivers and springs in the valleys). The Court held that there was no doubt that the quarrying operations negatively affected the water springs, which led to the encroachment of the right to life and personal liberty under Article 21 of the Constitution, and so allowed the petition under Article 32 of the Indian Constitution as a writ.

Following the implementation of CSR in South India, eight villages in the Guntur district benefited that were bearing the brunt of toxic releases from CCL Products (India) Ltd, a coffee exporting company, not only affecting human beings but property too. CSR can thus be beneficial for the protection of the environment. In the case of *Krishan Kant Singh v. National Ganga River Basin Authority,*[38] the tribunal held that CSR is not only a responsibility but the statutory obligation on companies to ensure there is no pollution. CSR may embody many different activities, and the above are merely examples. Individual businesses cannot choose which actions will be of more significant value to the entity or to society at large. In the case of *Additional Commissioner of Income Tax v. Rashtriya Ispat Nigam Ltd,*[39] it was held that expenditure incurred by the assessee company on repairs of roads, the supply of drinking water to villages, renovation/construction of a community center, literacy programs, health care, developmental work for development of peripheral villages where the persons who were displaced during the plant's construction, are presently staying as a part of the CSR.

Apart from the environmental implications, CSR may be a concern for society too. Relevant activities include: eradicating extreme hunger and poverty, the promotion of education, promoting gender equality and empowering women, reducing child mortality and improving maternal health, combating HIV/AIDS, malaria, and other diseases, ensuring environmental sustainability, enhancing vocational skills, and social business projects. Also included are contributions to the Prime Minister's National Relief Fund or any separate fund set up by central government or state governments for socio-economic development and relief and funds for the welfare of the Scheduled Castes, Scheduled Tribes, other deprived classes, minorities, and women. In *Additional Commissioner of Income Tax, Range 3 v. Rashtriya Ispat Nigam Ltd,*[40] it was found that CSR constituted activities without which industries are not able to function. In order to run an industry, it is the industry's responsibility to consider the environment and wider society, such as in the case of *M/S Kizhakethalackel Rocks v. Kerala State Level Environment Impact Assessment Authority and State of Kerala,*[41] wherein the appellant was engaged in mining activity that caused undue noise during blasting operations. The committee suggested conducting free medical checkups for people living within 300 meters of the project site, which can be performed once a year as part of CSR. Seen as a further aspect of CSR, in *National Aluminium Company Ltd and Ors. v. Ananta Kishore Rout,* it was held that schools should be set up by company NALCO acknowledging its responsibility as a model employer. NALCO wanted to provide this facility in the two NALCO campuses as a welfare measure. After supplying land, building, and infrastructure, and setting up the schools, everything was handed over to an outside agency to run them. An outside agency employed the staff and settled their service conditions.[42] In *Commissioner of Income Tax v. Infosys Technologies Ltd*,[43] the High Court held that to discharge their CSR, the company should incur the expense of installing traffic signals at Bannerghata Circle, a square near their office, which would also help their business if their employees were to reach work early as a result.

However, in the case of *Tata Power Company Limited v. Maharashtra Electricity Regulatory State Commission*,[44] it was observed that the expenses incurred toward CSR are in fact necessary for its electricity business. Further, research has determined that costs of CSR are passed on to consumers even though CSR is the social obligation of the corporate entity and as such should not be passed on to consumers. However, if the appellant's contention were to be accepted, then the appellant's customers would be paying for the discharge of the appellant's social responsibility. It is for the appellant to shoulder the burden of its CSR. Based upon such findings the Court held that the appellant activity was unjustified and so cannot be termed CSR.

8.5 Carbon trading and ecosystem

Plants create their food through photosynthesis: they absorb water from the ground and take in sunlight and carbon dioxide and oxygen is a by-product. However, excess carbon dioxide is harmful to the environment, and it may affect biodiversity. Industries are continually seeking ways ultimately to help the ecosystem to reduce excess carbon and maintain the balance of the environment. US and other international experts have supposedly reached a consensus to restore the changes made to the ecosystem by reducing carbon emissions and investing in research on alternative energy sources.[45] However, carbon emissions are tradable[46] and nowadays carbon credit (CR) is a very important factor in the industrial market. Ultimately, it can minimize tax for large corporations. There are two kinds of carbon credits: certified emission reductions (CERs) and verified emission reductions (VERs). In the case of certified emission reductions, the carbon credits are generated under the United Nations Framework Convention on Climate Change approved by a mechanism such as the clean development mechanism (CDM). However, in the case of verified emission reductions, the carbon credits are generated by the projects voluntarily agreed to within independent international standards. Both certificates—certified emission reductions and verified emission reductions—indicate the amount of the reduction of greenhouse gas emissions in the environment. Given this, it is evident that carbon credit is nothing but a measurement given to the levels of greenhouse gas emission rates in the atmosphere in the process of industrialization, manufacturing activity, etc. Therefore, a carbon credit is an entitlement given to industries for reducing greenhouse gas emissions in the course of their industrial activities.

In the cases of *Ambika Cotton Mills Ltd v. Deputy Commissioner of Income Tax*,[47] *My Home Power Ltd v. Deputy CIT*,[48] and *Sri Velayudhaswamy Spinning Mills P. Ltd v. Deputy Commissioner of Income Tax*,[49] the Hon'ble judicial and quasi-judicial body of India held that carbon credit is in the nature of "an entitlement" received to improve world atmosphere and environment-despoiling carbon, heat, and gas emissions. The entitlement earned for CR can be regarded as a capital receipt and cannot be taxed as a revenue receipt. It is neither generated nor created due to carrying on business, but rather accrued due to "world concern." It has been made available assuming the character of transferable right or entitlement only due to world concern. The source of carbon credit is thus world concern and the environment. Thus, the amount received for carbon credits has no element of profit or gain, and it cannot be subjected to tax in any manner under any head of income.[50] It is not liable for tax for the assessment year under consideration in terms of Sec. 2(24), 28, 45, and 56 of the Income Tax Act 1961. However, scientists consider it a by-product, so it is given to an assessee as a credit under the Kyoto Protocol and because of international understanding.[51]

Although the transfer of surplus loom hours to another mill is a capital receipt rather than income,[52] CR achieved high utility in the trading market by transferring certain considerations to other concerns and maintaining the environmental eco-balance by reducing the emission of harmful gases. A carbon credit is not like profit or income. It is not an offshoot of business but an offshoot of environmental concerns. The nature of its entitlement is to reduce carbon emissions; however, there is no cost of acquisition or cost of production to get this entitlement. In the *Commissioner of Income Tax v. Maheshwari Devi Jute Mills Ltd*,[53] it was observed that some 41 developed countries were committed to reducing greenhouse gas emissions by up to 5 percent during the period 2008 to 2012. Further, under the Kyoto Protocol, the quantity of greenhouse gas emissions does not bind less-developed countries: the Kyoto Protocol is binding on developed countries so as to meet the requisite target of carbon emissions reduction.[54]

A further facet of CR is "captive consumption": the assessee who generates electric power by using a gas turbine may also reduce carbon emission into the atmosphere. Since the United Nations Framework Convention on Climate Change implemented public CR, the assessee must have obtained the certified emission reduction/carbon credit in the course of its business activity, and it is a privilege or perquisite conferred on the assessee in the course of its manufacturing activity. In this case, the income on the sale of carbon credit is a trading/revenue receipt.[55]

8.6 Conclusion

This chapter is a touchstone to attempt a positive step forward in order to implement a better future with sustainable green development. This can be achieved through the scientific study of society, law, and economy determining the needs of civilized society. In this chapter, I have tried to connect all relevant factors of sustainable development with the intent of improving humanity both socially and economically. Ultimately, law should change things and promote the public good. Considering all the points discussed in this chapter simply shows that the law regarding the right to food or life and development needs to change.

8.7 Editors' note

In this chapter, Sumit Saurav provides a regional approach to agroecology and agrobiodiversity, describing the most important Asian trends in regulation reforms. He outlines some recent regulation developments in India, Bangladesh, Sri Lanka, and some other Asian countries, with specific reference to constitutional perspectives on the right to food. He also considers some more specific economic issues, like trade distortions in the market of basic food. The discussion on the right to food can be read in conjunction with Chapter 3 by Steier, where a more general approach is put into context with Sustainable Development Goals (SDGs).

A key issue in this chapter is corporate social responsibility, an economic concept with some legal implications which is examined with reference to trade and industry regulation, taking also into consideration international law sources.

After general remarks about environmental issues, Sumit Saurav focuses on specific challenges relating to food rights, with specific reference to public distributive systems (PDSs) and direct cash transfer systems (DCT) in India.

The author also applies the juridical and economical concept of corporate social responsibility to the new trends of Indian case law on human rights, with specific reference to health-related cases.

After presenting some other cases related to carbon trading, he concludes his chapter by pointing out the need for a deep change of food rights regulation. We share this idea, and remark that a law reform perspective in such an essential human right is a key issue in constitutional and international law which represents a challenge for legislators at the present time and in the near future.

Notes

1. Siddiq Osmani (ed.), *Choice, Welfare, and Development* (New York: Oxford University Press, 2001), p. 253.

2. Andrew Clapham, *Human Rights: A Very Short Introduction* (New York: Oxford University Press, 2007), p. 43.
3. *People's Union for Civil Liberties v. Union of India & Others*, writ petition (Civil) No. 196 of 2001.
4. Supreme Court Order dated October 29, 2002 in the writ petition (Civil) No. 196 of 2001.
5. Supreme Court Order dated May 8, 2002 in the writ petition (Civil) No. 196 of 2001.
6. Ibid.
7. Ibid.
8. Supreme Court Order dated April 27, 2004 in the writ petition (Civil) No. 196 of 2001.
9. Supreme Court Order dated September 17, 2001 in the writ petition (Civil) No. 196 of 2001.
10. Self-sufficiency in rice and food security: a South Asian perspective, Springer Link https://link.springer.com/article/10.1186/2048-7010-2-10, at p. 122, accessed April 3, 2018; https://lawexplores.com/food-education-health-housing-and-work/.Food, education, health, housing, and work, December 26, 2015. Law explorer website.
11. *Maneka Gandhi v. Union of India* AIR 1978 SC 597. The Supreme Court observed that it would include all these aspects which would make life meaningful, complete, and worth living. Similarly, in *Shantistar Builders v. Narayan Khimalal Totame* (1990) 1 SCC 520, the Supreme Court stated: "The right to life is guaranteed in any civilized society. That would take within its sweep the right to food"; https://lawexplores.com/food-education-health-housing-and-work/.
12. See https://lawexplores.com/food-education-health-housing-and-work/.
13. Ibid.
14. *Kishen Pattnayak & Another v. State of Orissa*, AIR 1989 SC 677.
15. *People's Union for Civil Liberties v. Union of India and Others*, writ petition (Civil) No. 96 of 2001.
16. See Lidija Knuth and Margret Vidar, *Constitutional and Legal Protection of the Right to Food around the World*, Food and Agriculture Organization of the United Nations, 2011, www.fao.org/docrep/016/ap554e/ap554e.pdf.
17. Supreme Court Order dated May 2, 2003 in the writ petition (Civil) No. 196 of 2001.
18. *Dr. Mohiuddin Farooque v. Bangladesh and Others* (No. 1), of July 1, 1996.
19. Article 15 of the Bangladesh Constitution: "It shall be a fundamental responsibility of the State to attain, through planned economic growth, a constant increase of productive forces and a steady improvement of the material and cultural standard of living of the people, with a view to securing to its citizens ... the provision of the basic necessities of life, including food, clothing, shelter."
20. Article 27(2) of the Sri Lanka Constitution: "The State is pledged to establish in Sri Lanka a democratic socialist society, the objectives of which include ... the realization by all citizens of an adequate standard of living for themselves and their families, including adequate food, clothing and housing."
21. Article 38(d) of the Pakistan Constitution: "The State shall provide basic necessities of life, such as food, clothing, housing, education and medical relief, for all such citizens, irrespective of sex, caste, creed or race, as are

permanently or temporarily unable to earn their livelihood on account of infirmity, sickness or unemployment."

22. *The Right to Food*, Food and Agriculture Organization of the United Nations, www.fao.org/righttofood/news-and-events/news-detail/en/c/124029/#.U8YccJSSwRw, accessed May 1, 2019; "Supreme Court of Nepal ruling on the Right to Food."
23. Ghose Bishwajit, Sajeeb Sarker, Marce-Amara Kpoghomou et al., "Self-sufficiency in rice and food security: a South Asian perspective," *Agriculture & Food Security* (December 2013), https://link.springer.com/article/10.1186/2048-7010-2-10, accessed April 3, 2018.
24. Maros Ivanic and Will Martin, "Ensuring food security," *Finance and Development* 45(4) (2008), www.imf.org/external/Pubs/FT/fandd/2008/12/ivanic.htm, accessed April 4, 2018; "Supreme Court of Nepal ruling on the Right to Food."
25. "Food, education, health, housing, and work." December 26, 2015, Law Explorer website, https://lawexplores.com/food-education-health-housing-and-work/.
26. Clapham, *Human Rights*, p. 119.
27. See *Economic and Political Weekly*, 48(1) (January 5, 2013).
28. See *Chronicle*, 24(7) (2013).
29. Op ed., "Explanatory note: why we oppose the rush to cash transfers and UID," The Hindu (December 31, 2012; updated September 22, 2016, www.thehindu.com/opinion/op-ed/Explanatory-note-why-we-oppose-the-rush-to-cash-transfers-and-UID/article12469558.ece.
30. United Nations, "Food" (2017), www.un.org/en/sections/issues-depth/food/index.html.
31. Clapham, *Human Rights*, p. 256.
32. What the United Nations Industrial Development Organization (UNIDO) defines as a "concept of management wherein Companies integrate considerations for human rights, societal, environmental and climate concerns, and combating corruption in their business activities interactions with their stakeholders."
33. R.C. Saxena, *Labour Problems and Social Welfare* (Meerut: K. Nath, 1981), p. 306.
34. Section 135 of the Companies Act, 2013 provides:
    (1) Every company having net worth of rupees five hundred crore or more, or turnover of rupees one thousand crore or more or a net profit of rupees five crore or more during any financial year shall constitute a Corporate Social Responsibility Committee of the Board consisting of three or more directors, out of which at least one director shall be an independent director.
    (2) The Board's report under sub-section (3) of section 134 shall disclose the composition of the Corporate Social Responsibility Committee.
    (3) The Corporate Social Responsibility Committee shall,—
        (a) formulate and recommend to the Board, a Corporate Social Responsibility Policy which shall indicate the activities to be undertaken by the company as specified in Schedule VII;
        (b) recommend the amount of expenditure to be incurred on the activities referred to in clause (*a*); and
        (c) monitor the Corporate Social Responsibility Policy of the company from time to time.

(4) The Board of every company referred to in sub-section (*1*) shall,—
(a) after taking into account the recommendations made by the Corporate Social Responsibility Committee, approve the Corporate Social Responsibility Policy for the company and disclose contents of such Policy in its report and also place it on the company's website, if any, in such manner as may be prescribed; and
(b) ensure that the activities as are included in Corporate Social Responsibility Policy of the company are undertaken by the company.

(5) The Board of every company referred to in sub-section (*1*), shall ensure that the company spends, in every financial year, at least two per cent. of the average net profits of the company made during the three immediately preceding financial years, in pursuance of its Corporate Social Responsibility Policy:

Provided that the company shall give preference to the local area and areas around it where it operates, for spending the amount earmarked for Corporate Social Responsibility activities:

Provided further that if the company fails to spend such amount, the Board shall, in its report made under clause (*o*) of sub-section (*3*) of section 134, specify the reasons for not spending the amount.

Explanation.—For the purposes of this section "average net profit" shall be calculated in accordance with the provisions of section 198.
Rule 3 of the CSR Policy, 2014 provides that

(1) every company including its subsidiary, and a foreign company under Section 2(42)of the Act, having its branch office or project office in India, which fulfils the criteria specified in Section 135(1) of the Act, shall comply with the provisions of Section 135 of the Act, and these Rules:

Provided that the net worth, turnover or net profit of a foreign company shall be computed in accordance with the balance sheet and profit and loss account of such company prepared in accordance with the provisions of Section 381 and Section 198 of the Act.

(2) Every company which ceases to be a company covered under Section 135(1) of the Act for three consecutive financial years shall not be required to—
(a) Constitute a CSR Committee;
(b) Comply with the provisions contained in sub sections (2) to (5) of Section 135;
(c) till such time it meets the criteria specified in Section 135(1).

35. P. Jayanth, "Toxic effluents impacting fields, public health," *Times of India* (June 1, 2015).
36. AIR 1980.
37. AIR 1985.
38. NGT, M.A. Nos. 879 of 2013 and 403 of 2014 in Original Application No. 299 of 2013. Decided on 16.10.2014.
39. ITA Nos. 09, 16, 14, and 15/Vizag/2013. Decided on 29.04.2015.
40. Osmani, *Choice, Welfare, and Development*.
41. National Green Tribunal, Appeal No. 29 of 2013. Decided on 13.02.2014.
42. Civil Appeal Nos. 5989, 5992, and 5993 of 2008. Decided on 08.05.2014.

43. (2014) 270 CTR (Kar) 523.
44. Appeal No. 104 of 2012. Decided on 28.11.2013.
45. Mathew Carr, "U.S. said to engage in talks on carbon market rules" (October 19, 2015), www.bloomberg.com/news/articles/2015-10-19/u-s-said-to-propose-fallback-global-emissions-trading-markets, accessed October 20, 2017.
46. *Apollo Tyres Ltd v. Assistant Commissioner of Income Tax*, 2014 (31) ITR (Trib) 477 (Cochin).
47. 2013 (27) ITR (Trib) 44 (Chennai).
48. 2013 (21) ITR (Trib) 186 (Hyd).
49. 2013 (27) ITR (Trib) 106 (Chennai).
50. Tax Article, Udyog Software, www.udyogsoftware.com/wp-content/uploads/2013/09/Article-11.06.2014-to-20.
51. *Tuticorin Alkali Chemicals and Fertilizers Ltd v. Commissioner of Income Tax*, 1997 (227) ITR 172 (SC).
52. *Commissioner of Income Tax v. Maheshwari Devi Jute Mills Ltd*, 1965 (57) ITR 36 (SC).
53. Ibid.
54. *Cadell Weaving Mill Co. P. Ltd v. Commissioner of Income Tax*, 2001 (249) ITR 265 (Bom.).
55. *Metal Rolling Works Pvt. Ltd v. Commissioner of Income Tax*, (1983) 142 ITR 170 (Bom.) and *O.K. Industries v. Commissioner of Income Tax*, (1987) 163 ITR 51 (Ker.).

chapter nine

Agrobiodiversity loss and the construction of regulatory frameworks for crop germplasm

Susannah Chapman and Paul J. Heald

Contents

> 94% of our seed diversity has disappeared.
>
> **(*Seed: The Untold Story 2018*)**
>
> Any story has a propensity to generate another story in the mind of its reader (or hearer), to repeat and displace some prior story.
>
> **(Clifford 1986: 100)**

9.1 Introduction

One of the major challenges facing agricultural and food systems today is the loss of crop diversity. Since at least the 1970s, a good deal of research has sought to track genetic erosion—or negative net change in the diversity of crop populations—alongside the factors that might support the maintenance of crop diversity. This empirical work has been complemented by an array of gray literature (FAO 1996, 2010; Global Alliance 2016), popular non-fiction (Fowler and Mooney 1990; Shiva 1993, 1997; Siebert 2011; Ray 2012), and film (*Seeds of Time* 2013; *Seed: The Untold Story* 2018), much of which has aimed to evaluate, explain, and raise awareness about agrobiodiversity loss and genetic erosion in crop populations. These accounts of erosion have linked the decline of crop diversity to broader processes of social and cultural change, including agricultural modernization and international development, the industrialization of food production, and the disappearance of "traditional" agrarian practice. Whether or not such accounts actually document instances of loss, they have gained resonance through the environmental, social, and political transformations they evoke. In this sense, efforts to explain and theorize the erosion of crop diversity—much like other accounts of environmental change—are also exercises in narration (Brush 2004; Montenegro de Wit 2016; Nazarea 2017).

This chapter explores the productive interplay between empirical accounts of loss and their narrative counterparts. We argue that the narrative dimensions of loss not only shape how changes in agrobiodiversity are explained and theorized, they also can influence the ways in which actual changes in agrobiodiversity are measured, described, and incorporated into programs to enhance diversity. Attention to the narrative dimension of loss is important because the stories relayed through accounts of genetic erosion affect the ways in which changes in crop diversity are interpreted and imported into projects to conserve and create diversity. For example, where accounts of loss foreground the abandonment and deterioration of crop diversity due to encroaching modernization, they tend to obscure the practices of everyday conservation that support *in situ* diversity. In rendering partial stories of diversity dynamics, loss narratives lend urgency to *ex situ* projects fashioned on models of salvage and containment (Brush 2004; Montenegro de Wit 2016; Nazarea 2017).

One focus in this chapter concerns a similar effect that loss narratives have on apprehending and explaining the proliferation of new diversity: We are particularly interested in how the longstanding narrative coupling of loss and agricultural modernization makes certain types of new diversity more difficult to accommodate in accounts of crop diversity. Where new diversity surfaces in empirical accounts, it may be treated as an outlier, attributed to the modern breeding programs that have produced genetically uniform cultivars, or not counted at all. This one-dimensional treatment of new diversity is problematic because it can obscure the ongoing, dynamic, and innovative work of farmers, gardeners, immigrants, and seed savers in the creation and distribution of new diversity. Somewhat ironically, in parsing out new diversity, accounts of erosion can gesture toward the very conditions of scarcity that undergird arguments for exclusive intellectual property as a necessary component of agricultural innovation: that there is both a lack of new and useful plant varieties and a dearth of inspiration to innovate.

Thus, while the loss of crop diversity is a concern to be taken seriously, we think that it is also crucial that attempts to narrate and describe that loss make space to acknowledge the processes at play that produce diversity. This is not a denial that crop diversity loss has happened or that it should not be taken seriously. Rather, it is an attempt to explore how loss, as it has been theorized and narrated, can nonetheless sideline, marginalize, and even act as a filter on otherwise well-documented cases of the persistence and proliferation of crop diversity in ways that are potentially consequential for efforts to conserve, distribute, and innovate diversity in the future.

In making our case, we revisit a previous study of ours that tracked changes in open-pollinated, commercially available vegetable and apple diversity alongside transformations in intellectual property law in twentieth-century USA (Heald and Chapman 2012).[1] In that study we found that while the twentieth century had witnessed great loss of historical vegetable diversity, there had been almost complete replacement of that varietal diversity from a number of sources: importation, non-intellectual-property-protected innovation, and, to a much lesser extent, intellectual-property-protected innovation of open-pollinated varieties. We reflect upon those findings in the light of two phenomena. The first is the traction of one of the most widely cited accounts of agrobiodiversity erosion, the findings of a 1983 Rural Advancement Fund International (RAFI) study that our study directly engaged. The second is the array of responses elicited by our study from the interpretation that we *must* have been tracking the innovation of hybrids and scientifically bred varieties (Ray 2012) to the curious fact that it is our numbers on loss—but not our numbers on varietal replacement—that have become the taglines about erosion for movies such as *Seed: The Untold Story* (2018).

9.2 *Theorizing genetic erosion*

Theories of genetic erosion first emerged in the early twentieth century. As early as 1914, the German agronomist Erwin Baur began to entertain the idea that genetic erosion might occur within cultivated plants and their wild relatives (Hammer and Teklu 2008). It was not until the 1930s, however, that concerns about the loss of agricultural diversity began to gain traction. In 1936, the American crop explorers H. Harlan and M. Martini sounded an alarm over the loss of ancient barley landraces. In their overview of barley breeding programs published in the US *Yearbook of Agriculture*, they wrote about crop diversity loss in the "hinterlands" of Asia:

> The progenies of these [barley] fields with all their surviving variations constitute the world's priceless reservoir of germ plasm. It has waited through long centuries. Unfortunately, from the breeder's

> standpoint, it is now being imperiled. Historically, the tribes of Asia have not been overfriendly. Trade and commerce of a sort have always existed. They have existed, however, on a scale so small that agriculture has been little affected. Modern communication is a real threat ... When new barleys replace those grown by the farmers of Ethiopia or Tibet, the world will have lost something irreplaceable.
>
> **(Harlan and Martini 1936: 317)**

Early theorists of genetic erosion were concerned that the spread and adoption of varieties bred by formal breeding programs would displace the landrace varieties grown within areas of historically diverse agricultural production. In the early twentieth century, the rise of line breeding practices in scientific plant breeding programs enabled revolutions in plant breeding, particularly through increases in yield. However, the varieties they produced were unique in their genetic uniformity. Alongside widespread enthusiasm about the promise of modern plant-breeding techniques, there was also a sense that the spread of the very varieties those techniques produced could cause loss by displacing farmers' more diverse cultivars (Brush 1999). This concern rested on a number of assumptions. The first was that in adopting formally bred varieties farmers would necessarily abandon their existing varietal repertoires (Brush 2004). This assumption extended a more general proposition about social and technological change: The idea that various "modernizing" forces—access to markets, changes in technology, long distance communication—would facilitate varietal diffusion, transforming traditional agricultural practice and, in the process, initiate the process of genetic erosion.

That such changes might occur within the hinterlands of the Global South seemed to recall the transformations of agricultural production experienced by industrialized nations in the nineteenth and at the turn of the twentieth centuries. The rise of modern plant-breeding practices and the growing prevalence of monoculture production in places like the USA seemed to foreshadow the state of agricultural diversity to come. Before the loss of crop diversity could be empirically documented, the perceived genetic erosion of food production in industrialized countries became testament to the future of agricultural production elsewhere. The concept of loss was thus entwined with broader ideologies of social progress and modernization, processes that were at the same time causes of genetic erosion and testament that it had happened.

The notions of progress and modernization that underlay concerns about genetic erosion came to shape the aims of international agricultural development initiatives that took hold in the middle of the century.

Ideas about "self-help" and the modernization of food production that developed within twentieth-century agricultural extension in the USA were exported—along with genetically uniform, high-yielding, and high-response varieties—through the early Green Revolution initiatives of the 1950s and 1960s (Nally and Taylor 2015). During the second half of the twentieth century, the modern varieties produced by Green Revolution breeding programs spread widely. Depending on crop species and world region, by the 1990s, modern varieties accounted for up to 90 percent of cultivated land area (Evenson and Gollin 2003). The rapid and widespread dissemination of these varieties throughout much of Latin America, Central Asia, and Southeast Asia under the auspices of the Green Revolution spurred a number of new actors to action in debates over genetic erosion (Mooney 1983; Pistorius 1997). Where agronomists and plant explorers had located the beginnings of genetic erosion in the earliest releases of formally bred varieties during the early twentieth century, by the late 1960s and early 1970s, others had begun to point to the dominant agricultural model of the mid-twentieth century as a driver of agrobiodiversity loss (Frankel 1988).

In the early 1970s, concerns about genetic erosion propelled efforts to measure, theorize, and raise awareness about the loss of crop diversity. Stephen Brush (2004) has noted that concern over the replacement of landraces and their wild relatives was solidified in the 1970 publication edited by Otto Frankel and Erna Bennet, *Genetic Resources in Plants.* Therein, Frankel and Bennet laid out the problem of both the narrowing of the genetic base used in formal breeding programs and the eventual displacement of farmers' varieties with those released by breeders. Amidst concern that "the treasuries of variation in the centres of genetic diversity will disappear without a trace," Frankel and Bennet put forth "an *a priori* case for urgent measures to protect and preserve valuable primitive and wild gene pools which are threatened with extinction" (1970: 9–10). Their project was nonetheless constrained by a lack of empirical evidence. Although the specter of erosion organized the theme of Frankel and Bennet's volume, Brush (2004) has pointed out that its contributions were marked by a conspicuous lack of data from case studies demonstrating actual instances of genetic erosion. Brush found a similar lacuna in a follow-up volume published by Frankel and Jack Hawkes in 1975, in which only three contributions presented data about genetic erosion within farmers' fields. These included Frankel's own overview of the 1970/71 FAO *Survey of Crop Genetic Resources in Their Centres of Diversity,* another chapter that failed to find evidence to support the genetic erosion hypothesis, and the volume's contribution by Ochoa, who confirmed a decline in the number of landrace potatoes cultivated in parts of Chile and Peru.

Nonetheless, in 1970, two events coincided to reinforce growing disquiet about genetic erosion and genetic uniformity in crop populations:

the spread of southern corn leaf blight in the USA and an outbreak of coffee rust in Brazil. Reminiscent of the Irish potato famine caused in part by the rapid spread of *Phytopthera infestans*, the southern corn leaf blight of 1970 wiped out much of the US corn crop, which had been bred to be cytoplasmically uniform (Harlan 1975). Likewise, Brazil's outbreak of coffee rust led to major losses in the crop, causing worldwide price fluctuations (Scarascia-Mugnozza and Perrino 2002). The massive losses to the US corn and Brazilian coffee crops were largely due to the vulnerability of both monoculture and the pure-line breeding practices that were used in many crop development programs at the time (van de Wouw et al. 2010). Although these events were indicative of serious genetic uniformity within these different agricultural landscapes, they were not measures of erosion of total diversity as envisioned in early theories of genetic erosion. Regardless, these cases served as powerful anecdotes about loss because of what they relayed about the dangers of the narrow genetic base of many commercial crops. In a very real way, they affirmed the concerns about the genetic uniformity of modern cultivars voiced by the earliest theorists of genetic erosion, even though they did not indicate an absolute loss.

Taken together, the 1970 survey of genetic erosion and a string of crop failures helped mobilize the FAO, the World Bank, and the UN Development Program to initiate projects to survey, collect, and bank germplasm found in farmers' fields. In 1971, the Consultative Group on International Agricultural Research (CGIAR) was established, and by 1972, in a meeting held in Beltsville, Maryland, attendees drafted a plan for the creation of a global network of Genetic Resource Centers that would specialize in the conservation of different crop species (Harlan 1975; Pistorius 1997). Soon thereafter, the International Board for Plant Genetic Resources was founded in order to promote and support the worldwide collection and conservation of plant germplasm for the sake of future research and production (Scarascia-Mugnozza and Perrino 2002). Describing the effect that emergent concerns over erosion narratives had on the direction of conservation policy, Brush wrote that "little [was] known about the actual crop populations in question, and even less about the farming cultures that produce[d] them. With this uncertainty, the safest choice [was] to assume the worst, that genetic erosion would shortly eliminate landraces" (1995: 346).

By the early 1970s, the concept of genetic erosion had found its place within international agricultural policy as well as the scholarly literature on crop diversity. The threat of genetic erosion within crop populations, however, increasingly spoke to concern over genetic scarcity, or, in the words of Pat Mooney, the "not-so-renewable renewable resource" (1983: 7). The idea was that locally adapted cultivars were precious and unique genetic material that, once lost, could never be gotten back. This was echoed in Jack Harlan's warning that:

> In view of the obvious limitations of our collections and in the face of the current genetic "wipe out" of centers of diversity, it may be too little and too late. We continue to act as though we could always replenish our supplies of genetic diversity. Such is not the case. The time is approaching, and may not be far off, when essentially all the genetic resources of our major crop will be found either in the crops being grown in the field or in our gene banks.
>
> **(Harlan 1975: 621)**

Over the next two decades, concerns about scarcity materialized in roughly two ways. In the first manifestation, genetic erosion became evidence of the technological superiority of high-modernist agricultural development (Brush 2005), a testament that if given the opportunity, farmers would make choices like economists predicted and select varieties based on a few narrow criteria, namely the increased yield achieved through scientific breeding programs. The erosion implied in this model was troubling because modern plant breeding depended upon farmers' landraces at the same time that the varieties breeding programs produced might be replacing the diversity of landraces in farmers' fields (Hardon 1996). The precariousness (and irony) of this situation lent urgency to calls for *ex situ* conservation—particularly over other models, such as *in situ* or *in vivo* conservation (Brush 1995; Hunn 1999)—at the same time that it reaffirmed the importance of crop diversity for the future of scientific plant breeding and varietal innovation.

In the second manifestation, concern over genetic erosion and its potential endgame, genetic scarcity, emerged as a trenchant critique of the political economy of twentieth-century food production and the growing corporate influence within the seed industry (Mooney 1983; Fowler and Mooney 1990). Transformations in twentieth-century food production—changes epitomized by Green Revolution policies—had created a scenario where "[t]he crop and varietal diversity of indigenous agriculture was replaced by a narrow genetic base and monocultures" (Shiva 2015: 31). Where this second positioning often gestured toward the value of *ex situ* conservation, it simultaneously raised questions about justice and equity in the emerging international system for plant governance and agricultural development, underscoring the need for conservation by many means. And, if the first assemblage of concerns around genetic erosion saw redemption in the promise of modern plant breeding, this second manifestation offered an incisive critique of that very enterprise and its increasingly entwined relations with private industry and intellectual property. In this critique, the standard of uniformity required for patent and plant variety protection was "yet another factor contributing to the narrowing of the genetic base of our crops" (Fowler and Mooney 1990: 87) all the while that it was through this very system of patenting that varietal "contributions of farmers go unrecognized, unprotected—even denigrated" (Fowler and Mooney 1990: 145).

In both narrative instantiations, the dangers of genetic uniformity and the changes wrought by dominant models of twentieth-century agriculture were taken as testament to the threat of genetic erosion. The narrow genetic footprint of formally bred varieties and a motley array of "modern" agricultural practices (commercial production, varietal adoption, industrialization, intellectual property) came to represent erosion, despite the fact that there was little long-term data on crop diversity dynamics, crop populations, or even the farmers who maintained them (Brush 1995). In one sense, the narratives that took shape in the process of identifying and explaining perceived changes in crop diversity can be taken as a form of necessary interpretation—a way of making sense of a set of quite complex socioecological transformations within the context of the twentieth century. In another sense, however, loss narratives also relayed a suite of additional stories about the conditions of modernity, the loss of "traditional" lifeways, and the orderings of social transformation (Brush 2005). Like the allegorical register of ethnography (Clifford 1986), the narrative power of loss is as much an interpretation of genetic erosion as it is a condition of erosion's meaningfulness. While this narrative power necessarily shapes the ways in which changes in crop diversity are explained, it also influences how those changes are measured, described, and interpreted.

As we see it, this narrative power can intervene in the measurement and description of crop diversity in at least two ways. The first—and perhaps the most common—tendency is to frame accounts of diversity persistence against a general backdrop of crop diversity loss. Maywa Montenegro de Wit (2016) has observed this tendency at play in instances where examples of robust, contemporary *in situ* diversity are positioned as outliers or exceptions to an otherwise obvious downward trend in crop diversity. While Montenegro de Wit points to a number of scholarly works that have positioned evidence of diversity maintenance in this way, this framing is also a common tactic used in popular accounts of loss. Take, for example, Charles Siebert's 2011 *National Geographic* article on genetic erosion. It begins with a vibrant description of the idiosyncratic diversity of the Seed Savers Exchange farm in Decorah, Iowa, set against the monoculture of the neighboring fields. It ends with a more sobering account of the invaluable seed diversity cultivated by one Ethiopian farmer, a visit to whose "land is like tumbling back in time to an ancient way of farming." Set between these two outliers of diversity (notably, Seed Savers is an outlier in space, the Ethiopian farmer in a narratively constructed time) is a stark rendering of the crop genetic erosion of the twentieth century. Despite these examples of persistence, Siebert notes that "food varieties' extinction is happening all over the world—and it's happening fast." Citing what appears to be updated figures (Heald and Chapman 2009a) on RAFI's 1983 report of US crop diversity, he explains that "an estimated 90 percent of our historical fruit and vegetable varieties have vanished ... It took more than 10,000 years of domestication for humans to create the vast biodiversity in our food supply that we're now watching ebb away." Loss, as part and parcel of the triumph of modern agriculture—if not the modern condition—is happening, even when and where crop diversity persists.

The second influence of loss narrative power has come to bear on the types of data taken into measures of erosion, such as the way in which baselines get drawn and the types of diversity that get counted against those lines. Although sometimes decisions about what to measure are limited by the available data, some of the most widely cited figures of crop diversity loss track only historical varietal diversity into the present context, thereby excluding diversity that emerged after the study's baseline was drawn. To be clear, there is a large amount of empirical work on crop diversity that has *not* tracked diversity in this way, focusing instead on varietal dynamics and farmer management strategies (Zimmerer 1997; Brush 2004), counts of varietal richness (Jarvis et al. 2008), or measures of total inter-varietal and intra-varietal allelic diversity at various times (Fu et al. 2003; Fu and Somers 2009; Bonneuil et al. 2012). Yet, in narratives of loss, it is a common tactic to track that historical diversity into the present to the exclusion of other counts of diversity.

One of the common rationales for tracking changes in only historical diversity is that "newer" diversity may have origins in the very pure line breeding programs that sparked concerns of genetic erosion. Where this is the case, measures of varietal diversity may actually elide genetic uniformity, making new diversity a poor if not absolutely inappropriate equivalent for old diversity (Guarino 2012). Such a concern reflects the very real problem of genetic uniformity in twentieth-century plant breeding practices and the additional confounding dilemma of having to use varietal counts as proxies for measures of genetic diversity. Indeed, a recent analysis tracking genetic and varietal diversity in French wheat from the late nineteenth century to the early twenty-first century has found that increases in richness of varietal diversity can mask declines in genetic diversity (Bonneuil et al. 2012). More varieties, in other words, can still reflect reduced diversity, particularly in cases where the varieties in question share a similar genetic structure, where a handful of varieties become dominant across the landscape, or when more diverse landraces are replaced by more uniform cultivars (Bonneuil et al. 2012).

At the same time, tracking only the loss of historical diversity can produce elisions of its own. In cases where historical genetic data may not be available or where counts of overall varietal richness may be one of the only measures available for assessing diversity, should it be assumed that all new diversity represents that which is more uniform and, by extension, is at some point the output of modern breeding programs? Certainly, in the absence of genetic analysis, caution should be taken not to extrapolate between varietal and genetic diversity. Similar caution should be taken in assessing the sources of new diversity. Yet, to draw a baseline of diversity against which only losses are calculated is more than a decision about method: In a world where some change is inevitable, tracing only historical diversity will inevitably produce declensionist results. By design, it enacts a disappearing, diverse past set against a uniform and increasingly diversity-scarce present. In the process, it also elides the contributions to new diversity made by all the other actors outside the remit of twentieth-century breeding programs.

With these influences on the measurement and description of crop diversity in mind, it is to the productive interplay between empirical measures and narrative accounts of loss that we now turn.

9.3 *Measuring genetic erosion*

In general, empirical measures of loss have been complicated by a lack of historical baseline data on crop diversity. Nonetheless, the net change in crop diversity—or vulnerability to it—has been assessed in a number of ways. The most common strategy for tracking net change has been to use varietal counts as proxy for change at the genetic level (Hammer et al. 1996; Hammer and Laghetti 2005), although measures of diversity at the allelic level are becoming more common (see Khlestkina et al. 2004; Fu et al. 2003; Fu and Somers 2009; van de Wouw et al. 2010; Bonneuil et al. 2012). Likewise, researchers have developed a number of tools to track crop diversity across space and time and thereby better differentiate between problems of uniformity, loss, and distribution. These include such measures as the "diversity count"—a means of gauging the total inter- and intra-specific diversity or "richness" of a production system—as well as the "diversity index measurement"—an index of the "evenness" or distribution of diversity within a specific locale (Eyzaguirre and Dennis 2007; Love and Spanner 2007; Jarvis et al. 2008).

Empirical measures of changes in intra-specific diversity richness have returned varied results. Hammer et al. (1996) traced changes in landrace varieties in Albania from 1941 to 1993 and in South Italy from 1950 to the 1980s. The authors reported 72.4 percent erosion of landraces across eight species in Albania and 72.8 percent erosion across thirty-two species in Italy. Hammer and Laghetti (2005) traced erosion rates of garden vegetables, pulses, and cereals in Italy since the 1920s. They reported varying rates of erosion throughout the twentieth century, finding the highest rates to have occurred in the 1930s. In other cases, studies measuring genetic diversity over time have reported allele flow between varieties without an overall change in the genetic diversity of collected specimens (Khlestkina et al. 2004); others have found qualitative changes in certain pools of wheat genetic diversity without overall quantitative changes at the genetic level (Donini et al. 2000); while others have charted overall declines in pools of wheat and oat diversity in Canada (Fu et al. 2003; Fu and Somers 2009). Likewise, an analysis of 44 published papers on genetic diversity trends for 8 different crop species released by plant breeders during the twentieth century found no substantial reduction of diversity (van de Wouw et al. 2010), while a 2012 analysis of wheat diversity in France reported that a proliferation of varietal diversity in wheat masked declines in diversity at the genetic level (Bonneuil et al. 2012). Such divergences and complexities in the empirical data on crop diversity erosion underscore how factors such as locale, decade, crop species, and even breeding program can influence the state of diversity. They also point to the slipperiness of the concept of diversity itself and some of the conceptual and practical difficulties that have arisen in attempts to measure it.

Research among farmers and home gardeners has further complicated assessments of crop diversity and accounts of crop genetic erosion. Specifically, research with farmers and home gardeners has documented the persistence of varietal diversity amidst the adoption of formally bred varieties, increased market integration, shifts in livelihood strategies, and changes in "traditional" practice (Brush 1992, 2004; Bellon 1996; Zimmerer 1997; Nazarea 1998; Veteto 2007; Jarvis et al. 2008). Such findings are significant because they demonstrate the persistence of diversity despite changes in production practices and economic relations that are so often linked to processes of genetic erosion. Likewise, research on farmer seed management has begun to suggest that farmers may regularly change their varietal repertoires as part of varied agronomic strategies (Brush 1999; Nuijten 2005; Teeken et al. 2012). In these instances, acts of varietal abandonment and adoption cannot easily be read into trajectories of loss because such turnovers suggest that baselines of diversity are themselves quite variable. Finally, a growing body of work in Europe and North America has documented the persistence of heirloom diversity among home gardeners, farmers, and seed savers, in large part because of the affective, sensory, and personal attachments the people have with the plants they cultivate (Nazarea 2005; Veteto 2007; Chapman and Brown 2013).

Yet, despite the varied snapshots of crop diversity provided by these empirical studies, accounts of widespread, general loss persist. Many of the accounts of major twentieth-century erosion are traceable to a survey of twentieth-century US apple, pear, and vegetable diversity conducted by RAFI in 1983 (see Chiosso 1983). The findings of that study (hereafter RAFI study), which were detailed in Cary Fowler and Pat Mooney's book *Shattering* (1990), have been widely credited as one of the most comprehensive measures of crop diversity loss during the twentieth century. For example, it was cited in the first FAO *State of the World* report (1996) as one of the best measures of longitudinal change in crop diversity. Likewise, Charles Siebert's (2011) *National Geographic* article on genetic erosion, and the films *Seeds of Time* (2013) and *Seed: The Untold Story* (2018), reference the RAFI study even though both actually present a slightly different set of data, a point to which we will return.

Undoubtedly, one reason that the RAFI study has been so influential is its historical and geographic scope. In many parts of the world, information about historical crop diversity may not exist, making comparisons between past and present diversity difficult or impossible. Sometimes data on historical diversity exists only as accessions entered into gene banks or as observational lists made by collectors on past excursions. Where such records do exist, they tend to provide data for changes in crop diversity in a specific locale within the relatively recent past (Hammer et al. 1996; Hammer and Laghetti 2005). Similarly, studies of farmer seed management are often confined to measures of diversity in a specific locale over a relatively short period of time. The RAFI study, on the other hand, compared the varieties listed in a number early twentieth-century United States Department of Agriculture (USDA) publications with the 1983 lists of varieties available at the US National Seed Storage Laboratory (NSSL). The turn-of-the-century USDA listings included all known commercially available varieties at that time, while the NSSL listings included those varieties that had been deposited within the seed bank. Thus, what set the RAFI study apart was its apparent ability to track nationwide varietal diversity for a large number of crop species over the course of almost a century.

Another reason the RAFI study has carried such weight is likely due to its rather stark findings. According to the study, 97 percent of the vegetable varieties listed in USDA compendiums at the turn of the twentieth century had been lost within eighty years (quoted in Fowler and Mooney 1990: 63).[2] For apples, the study reported 86.2 percent loss and for pears it reported 87.7 percent loss. When Fowler and Mooney invoked the study in their argument for *Shattering*, they were quick to point out a number of factors that might affect these numbers. First, like all studies of genetic erosion that rely on old varietal listings, there was a problem of sampling bias. In the case of the RAFI study, the 1903 compendium tracked only commercial diversity, meaning there were likely historical crop varieties that were not captured in the analysis. Second, they also pointed out that the RAFI study only compared diversity at the varietal level, which was an imperfect proxy for tracking genetic diversity. While some genes of lost varieties may persist elsewhere, they concluded that "given the magnitude of variety loss, one might even argue that many distinct genes and characteristics have been lost" (Fowler and Mooney 1990: 62–63). Finally, they noted that at least some of the lost varieties might persist in foreign seed banks and among "dedicated farmers or gardeners" (62). Whatever the shortcomings of the data presented by the RAFI study, Fowler and Mooney concluded that if the "study of the U.S. is any indication, extinction rates in other countries during the last century may be even more awesome. Yet day by day, the losses climb. More and more varieties become extinct, never to be seen again" (63).

Fowler and Mooney drew upon RAFI's findings to build their broader argument about the loss of crop diversity amid changing agricultural practices, Green Revolution policies, and the shifts in the governance of crop germplasm during the twentieth century. By tethering these concerns to RAFI's figures on genetic erosion, they crafted a critique of the political economy of food production that simultaneously invoked early theories of genetic erosion and extended concerns about agricultural "modernization" and its trappings (uniformity in breeding programs, development practice, changing traditional practice) to include a new suite of issues: privatization within the seed industry, new biotechnologies, conservation and collection initiatives, and the expansion of intellectual property law. *Shattering* was not necessarily the first nor was it the last to draw these connections (Mooney 1980; Global Alliance 2016). Part of the power of erosion narratives has been their ability to absorb all sorts of new culprits under the rubric of "modernizing" forces, and intellectual property has been foremost among the new suite of factors implicated in the loss of crop diversity. However, this narrative linkage, as we have shown elsewhere, can give culprits such as intellectual property far more credit than they deserve (for both erosion and innovation). What we want to bring to the reader's attention here is that this particular aspect of erosion narratives also has consequences for how we acknowledge "modern" (read "new") diversity and, by extension, those who might deserve credit for it.

9.4 *Twentieth-century loss, revisited*

We first came to the question of genetic erosion through a project to assess the effects of intellectual property protection on crop diversity in the USA during the twentieth century. Fully aware of Fowler and Mooney's book and the RAFI study, we were curious to explore two competing claims. The first claim that piqued our interest was that the application of intellectual property to crop varieties in the USA had led to the erosion of crop diversity. The second was that intellectual property rights were and are integral to the innovation of new diversity. We thus set out to track crop varietal diversity across major transitions in intellectual property law during the twentieth century. While generally wary of the amount of credit owed to intellectual property for either erosion or innovation, we nonetheless were of the understanding that the twentieth century had witnessed major declines in US vegetable and apple diversity.

For our study, we needed updated data for US crop diversity to cover the 25 years since RAFI's data. Because the 1903 USDA inventory used in the RAFI study surveyed commercial availability of seed catalogs, we decided to measure open-pollinated varieties available in 2004 commercial seed catalogs for 42 different vegetable species. To do this we compared the 1903 USDA listings used by RAFI with the US listings in the *Garden Seed Inventory* (Whealy 2005), a compendium of all open-pollinated, commercially available varieties in the USA and Canada. We carried out a similar assessment for apples by comparing the varietal listings in the *Fruit, Berry and Nut Inventory* (Whealy 2001) with historical nursery catalogs housed at the National Agricultural Library in Beltsville, Maryland. Both *Inventory* books are compiled by Kent Whealy, one of the founders of the Seed Savers Exchange, an organization dedicated to saving and exchanging varieties of heirloom, rare, and open-pollinated crops. Here we revisit some of the key empirical findings presented in our earlier study (Heald and Chapman 2012) so that we might better reflect on the narrative dimensions of agrobiodiversity erosion and the question of varietal innovation during the twentieth century.

A complete review of the methods we used are beyond the scope of this chapter. However, a few details should be mentioned. First, like other studies, we only tracked documented diversity, making it likely that there is varietal diversity not captured in our analysis. Our data did suggest that undocumented diversity is there in the margins, sometimes entering, sometimes leaving again the official ranks of the historical record. Second, we only traced changes in varietal counts, not changes in genetic diversity. The nature of the historical data precludes comparisons at the genetic level. Therefore, we cannot extrapolate our findings of loss or replacement to changes in genetic diversity. Finally, we were not measuring changes in the evenness or distribution of diversity. Given the prevalence of monocultures today, this is an aspect of US farm production that is undeniably important for theorizing loss and vulnerability to it. However, that measure was outside the scope of our study. Rather, what our findings provided was as a national-level study of the richness of commercially available diversity over the course of a hundred years.

Our data from the 2012 study revealed significant loss of historical diversity during the twentieth century as reflected in commercial listings. Of the approximately 7,262 varieties of 42 vegetable species available in 1903, only 6 percent were available in commercial catalogs in 2004. In other words, 94 percent of varieties available in the early twentieth century have been lost. This is a similar number to that documented by RAFI for 1983. Although RAFI and Fowler and Mooney reported 3 percent survival, we found in our research their numbers—due to a now-admitted calculation error—actually documented closer to a 7 percent survival rate.[3]

Despite this loss—and contrary to simple narratives of diversity erosion—our data also revealed extensive replacement of open-pollinated diversity. For example, in 2004, across the same forty-two vegetable species, we found that there were 7,100 open-pollinated varieties available through commercial sources. For apples, the persistence of old diversity and the emergence of new diversity has been even higher. Through our research, we estimated that between 280 and 420 varieties of apple were commercially available in 1905. Today, there are a total of 1,476 apple varieties available from commercial nursery catalogs. Additionally, approximately 75 percent of the apple varieties commercially available in 1905 were commercially available in 2000. In addition to the varieties available in commercial nursery catalogs, the Plant Genetic Resources Unit of the USDA maintains an additional 980 varieties, while the collections of private apple variety collectors contain even more.

We found that not all crop species experienced the same degree of varietal replacement (see Figure 9.1) and that the source of varietal replacement varied by crop (see Figure 9.2). Different species not only experienced differential degrees of erosion, but species such as garden beans, tomatoes, and garlic witnessed a proliferation of new varieties while species such as cabbage, sugar beet, and rutabaga experienced overall declines in diversity. These differential rates of loss and replacement

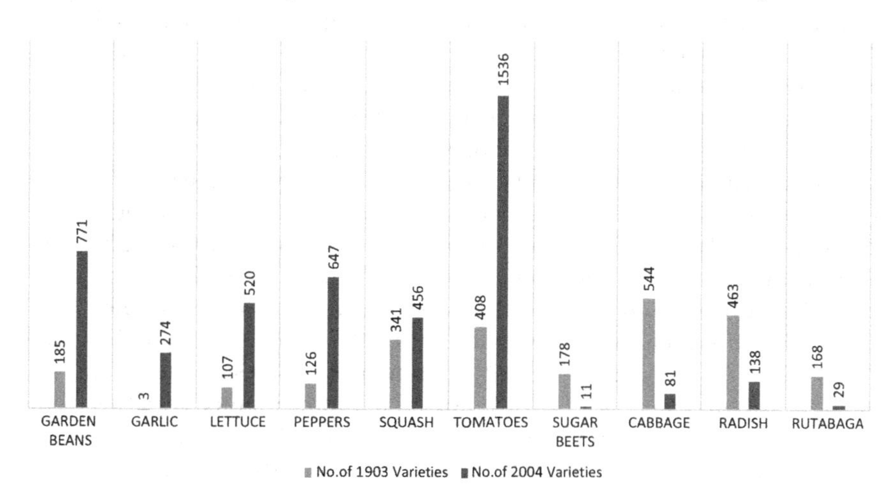

Figure 9.1 Crop species with greatest loss and replacement during twentieth century.

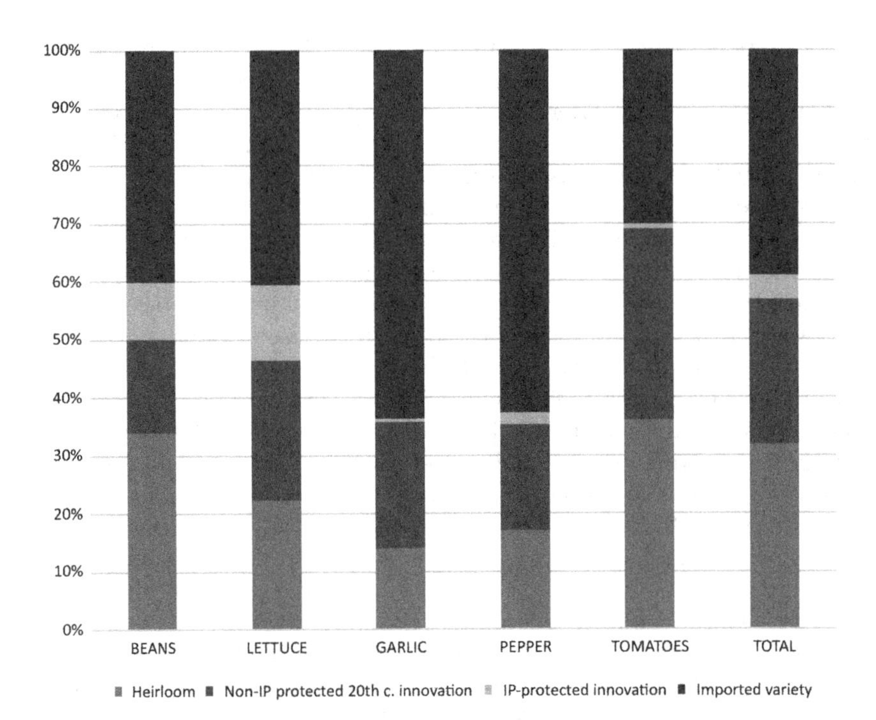

Figure 9.2 Sources of replacement diversity for select species.

are likely due to a number of factors including the reproductive biology of the plant, changing food habits, and transformations internal to US agricultural and seed industries. What these variable numbers suggest is that the history of open-pollinated vegetable and apple diversity in the US is dynamic and that various pools of diversity may be constantly in flux (Heald and Chapman 2012). A clearer picture of this dynamic emerges when we looked at the sources of new diversity by crop.

For all vegetable varieties available in 2004, we were able to determine that approximately 32 percent were of historical or "heirloom" origin, 39 percent were twentieth-century imports, 25 percent were non-patented American innovations, and 4 percent were patented innovations. Figure 9.2 provides a breakdown for sources of varietal diversity for five of the major vegetable crops included in the study. Figure 9.3 compares apple diversity in 1905 and 2000 and provides a breakdown of sources of varietal diversity in the year 2000. Significantly, both the importation of varieties from other parts of the world and localized (US) innovation of vegetables and apples constituted important sources of new diversity for vegetables and apples, while patented innovations contributed a much smaller proportion to the pool of available diversity.

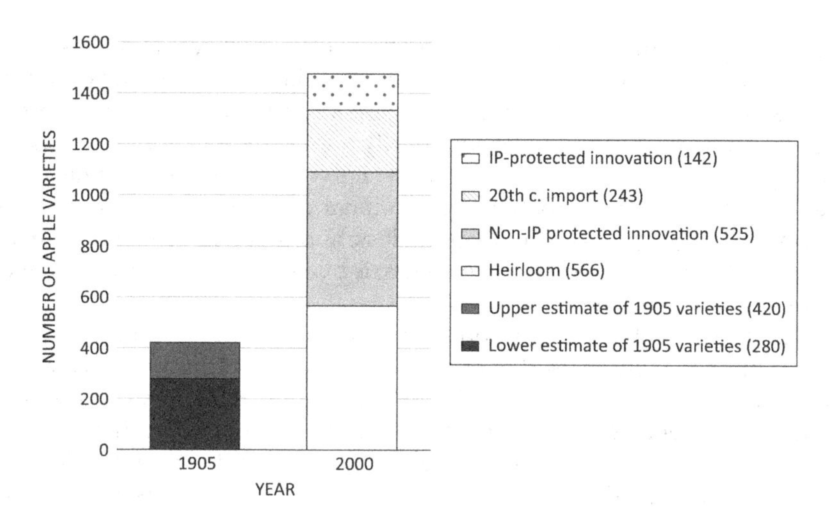

Figure 9.3 Commercially available apple varieties by year and source.

These findings are consistent with ethnographic and historical studies on agrobiodiversity in the USA that have demonstrated the contributions to crop diversity made by immigrants, indigenous peoples, seed savers, and home gardeners (Nabhan 1989; Airriess and Clawson 1994; Corlett et al. 2003; Nazarea 2005; Imbruce 2007; Veteto 2007). Taken together, the picture that emerges from these studies is also one far more complex than conventional accounts of loss communicate. Rather than simply pointing to extensive genetic erosion, this history of varietal diversity raises all sorts of questions about crop diversity dynamics: the incidence of varietal turnover; the special place of immigration in the maintenance of crop diversity; the relative value of the circulation of plant materials for the maintenance of diversity; the ability of heirloom or historical diversity to "hide out" in unexpected places; the impressive contributions of local, non-patented innovation to the creation of new diversity; the comparatively very slim contributions of intellectual property protection to the pool of open-pollinated diversity; even the way that and concerns about loss may have motivated efforts to stem it.

At times, the findings from our study have been read and circulated in ways that have largely forestalled these questions. This may be due partly to the ways in which conventional narratives of loss and modernization have shaped the lenses used to evaluate new diversity. In the circulation of our findings, we have noticed a tendency to read new diversity as if it were almost solely the product of the formal plant breeding projects that ignited concerns over genetic erosion. For example, in her poignant and beautiful account of crop diversity and seed saving in the USA, Janisse Ray cited a preliminary research report (Heald and Chapman 2009) where we first outlined major trends in loss and replacement. Acknowledging our updated figures on loss, she nonetheless raised doubts about the new diversity, noting that it was the product of "scientifically bred varieties" and thus was an incomplete replacement to historical diversity (2012: 6).[4] Lurking here are the major themes that structure conventional narratives of erosion: modernization has propelled crop diversity loss, and that which has come in its stead, by its sheer contemporaneousness, is by necessity substantively and problematically different.

Another way that conventional narratives of loss and modernization have shaped the lenses used to evaluate new diversity has been to cut out the data on innovation entirely. Although popular interpretations of our findings have been quite varied—with some reading the findings as evidence that there was little to be concerned about crop diversity loss to others who have offered more cautious but optimistic appraisals about what we should make of varietal replacement—it is our data on loss, not replacement, that has circulated the most widely.[5] For example, our updated figures for RAFI's 1983 study (seven versus three percent survival) were used in the film *Seeds of Time* to demonstrate the severity of twentieth-century erosion. Similarly, the tagline for the film *Seed: The Untold Story* invokes the finding from our 2012 study that 94 percent of seed diversity during the twentieth century was lost.[6] While it could be argued that the parsing of data on loss from data on replacement is merely a rhetorical tactic used to favor a particular argument about crop diversity, such a position runs the risk of focusing too much on the production or framing of any single telling of crop diversity loss. Indeed, comments such as "There cannot be as much diversity today! There cannot!" (which is a response from a farmer heard by one of the authors after a presentation of the data from the 2012 study) suggests that there is something of erosion narratives that has made new diversity hard to *think*.

9.5 Narrative effects

In this chapter, we have tracked some of the narrative dimensions of genetic erosion. Specifically, we have focused on how such narratives have come to interact with empirical accounts of crop diversity. Narratives of loss, we have argued, affect how accounts of diversity persistence are framed against general trends of erosion; they influence how the accounting of new diversity is treated in certain measurements of varietal change; and they shape how evidence of new diversity is interpreted within the history of scientific plant breeding.

To be clear, we are not questioning whether or not genetic erosion is an issue to be taken seriously (it should be) nor are we arguing that efforts to conserve crop diversity are unimportant. Indeed, even where extant diversity may be robust, the loss of historical varieties can be consequential for both biological diversity and cultural practice. Perhaps one reason that accounts of loss have been so harrowing and evocative is because of the affective attachments that humans have with the plants they cultivate. It is those very attachments that lend urgency to accounts of erosion and can propel the narrative dimensions of loss. However, what we do question are the ways in which erosion narratives draw simplistic equations between modernization and loss and, by extension, tradition and diversity. We also question the ways in which loss narratives invoke visions of scarcity and, in so doing, can simultaneously elide the production of new diversity and influence how that new diversity, when recognized, is interpreted (as the product of modern breeding programs).

It is these very elisions that are consequential for how efforts to conserve and create crop diversity are carried into the future. Where images of disappearing diversity favor the salvage and containment of landraces and farmers' varieties, images of scarce innovation favor policies that can supposedly spur the creation of new diversity. The irony is that in failing to acknowledge acts of hidden conservation, or the importance of distributed innovation, we may end up with policies that fail to take stock of the diversity of ways that humans engage in the innovation and distribution of new diversity. In other words, where some conventional erosion narratives point to intellectual property as a culprit in loss, they not only give intellectual property too much credit for destruction, they may enable intellectual property to take too much credit for innovation. The challenge ahead lies in how to discuss changes in agrobiodiversity without only resorting to narratives of scarcity. This is both an issue of being attuned to the additional stories told by loss—to erosion narratives' allegorical register—and a call to explore "replacement" diversity further as it emerges and circulates within seed economies.

9.6 Editors' note

Genetic erosion in the agroecology context is defined by the FAO as "the normal addition and disappearance of genetic variability in a population [that] is altered so that the net change in diversity is negative."[7] Simply put, the genetically modified or conventionally bred plant varieties are displacing the diversity of crops, thereby diminishing the overall pool of genetic material.

A hypothetical example may describe some varieties of rice available, which are replaced in the marketplace by only a handful of the most frequently traded ones. These most readily available varieties, however, may neither be the most nutritious nor the most popular with farmers and consumers, they are, however, likely the most carefully patented and commercialized ones that are subjects of key players in the globalized market.

The FAO summarizes that

> Some 10 000 different plant species have been used by humans for food and fodder production since the dawn of agriculture 10 000 years ago.
>
> Yet today just 150 crops feed most human beings on the planet, and just 12 crops provide 80% of food energy—wheat, rice, maize and potato alone provide 60%.[8]

This commercialization of plant genetic material, especially as part of the GMO debate, has been much publicized.[9]

The environmental impact of genetic erosion in agriculture has devastating effects, both for food system resilience and nutrition profiles around the world. With the reduction of food to just a few patented crops, a trend that has been worsening over the past three decades, people's diets are shifting toward the commodity crops, often highly processed and fattening, rich in simple sugars and carbohydrates, but low in nutrient density. The global pandemics of obesity, malnutrition, diabetes, and heart disease pay tribute to these sad developments, which find only some resistance in the so-called "food movement," which is largely beyond the scope of this volume.

The authors here in Chapter 9 provide a detailed account of the implications and the discourse centering around genetic erosion and present outstanding reflections on the context.

Notes

1. The findings of the 2012 publication were outlined in three previous research reports: one on vegetable diversity (Heald and Chapman 2009a), another on the role of patents and plant variety protection in the innovation of open-pollinated vegetable varieties (Heald and Chapman 2009b), and another on apple diversity and sources of innovation in the market for apples (Heald and Chapman 2010). We mention the timing of these reports because they influenced the ability of our findings to circulate prior to the publication of the final article (Heald and Chapman 2012).
2. This was calculated by tracking which varieties listed in early twentieth-century USDA compendiums were extant within the NSSL.
3. Noting this discrepancy is significant for the discussion at hand because it makes it possible to track how our findings were picked up, albeit only partially, within accounts of erosion.
4. One of the authors (Chapman) discussed this passage with Ray, after which she updated her assessment in subsequent printings.
5. For examples of more optimistic but cautious readings of the study, see Tribe (2012) and Guarino (2012). For readings that use our findings to imply that there is little to be concerned about crop diversity loss, see James Cooper's (2016) response to "Seed: the untold story." His blog cites our findings to argue that "all available research indicates that there is *not* a decrease in crop diversity" (his emphasis).

6. It is perhaps worth noting that the film's online educational materials attribute this finding to RAFI's 1983 study (*Seed: The Untold Story* 2018: www.seedthemovie.com/guide).
7. FAO, Genetic Erosion of Crop Populations in Centers of Diversity: A Revision (Stephen Brush, University of California, Davis), http://www.fao.org/wiews-archive/Prague/Paper5.jsp.
8. FAO, Erosion of plant genetic diversity, http://www.fao.org/newsroom/en/focus/2004/51102/article_51107en.html
9. See generally, Sheldon Krimsky and Jeremy Gruber, The GMO Deception (Skyhorse 2016).

References

Airriess, Christopher and David Clawson. 1994. Vietnamese Market Gardens in New Orleans. *Geographical Review* 84(1): 16–31.

Bellon, Mauricio 1996. The Dynamics of Crop Infraspecific Biodiversity: A conceptual framework at the farmer level. *Economic Botany* 50(1): 26–39.

Brush, Stephen. 1992. Ethnoecology, Biodiversity, and Modernization in Andean Potato Agriculture. *Journal of Ethnobiology* 12(2): 161–185.

Brush, Stephen. 1995. In Situ Conservation of Landraces in Centers of Crop Diversity. *Crop Science* 35: 346–354.

Brush, Stephen. 1999. Genetic Erosion of Crop Populations in Centers of Diversity: A revision. *Proceedings of the Technical Meeting on the Methodology of the FAO World Information and Early Warning System on Plant Genetic Resources.* Prague: Research Institute of Crop Production.

Brush, Stephen. 2004. *Farmers' Bounty: Locating crop diversity in the contemporary world.* New Haven: Yale University Press.

Brush, Stephen. 2005. Cultural Research on the Origin and Maintenance of Agricultural Biodiversity. In *Nature Knowledge: Ethnoscience, Cognition, Utility,* ed. Glauco Sanga and Gherardo Ortalli, 379–385. New York: Berghahn Books.

Bonneuil, Christophe, Robin Goffaux, Isabelle Bonnin et al. 2012. A New Integrative Indicator to Assess Crop Genetic Diversity. *Ecological Indicators* 23: 280–289.

Chapman, Susannah and Tom Brown. 2013. Apples of Their Eyes: Apple Trees and Memory Keepers of the American South. In *Seeds of Resistance/Seeds of Hope: Place and Agency in the Conservation of Biodiversity,* ed. Virginia Nazarea, Robert Rhoades, and Jenna Andrews-Swann, 42–64. Tucson: University of Arizona Press.

Chiosso, Elaine. 1983. *Vegetable Variety Inventory: Varieties from USDA 1903 list of American vegetables in storage at the National Seed Storage Laboratory.* Pittsboro: Rural Advancement Fund.

Clifford, James. 1986. On Ethnographic Allegory. In *Writing Culture: The poetics and politics of ethnography,* ed. James Clifford and George E. Marcus, 98–121. Berkeley: University of California Press.

Cooper, James. 2016. "Seed Diversity is Not a Serious Concern. Ignore the seed movie?" Food Science Institute. https://foodscienceinstitute.com/2016/09/16/seed-diversity-is-not-a-serious-concern-ignore-the-seed-movie/ (accessed March 15, 2018).

Corlett, Jan, Ellen Dean, and Louis Grivetti. 2003. Hmong Gardens: Botanical diversity in an urban setting. *Economic Botany* 57(3): 365–379.

Donini, P., J.R. Law, R.M.D. Koebner, J.C. Reeves, and R.J. Cooke. 2000. Temporal Trends in the Diversity of UK Wheat. *Theoretical and Applied Genetics* 100: 912–917.

Evenson, Robert and D. Gollin. 2003. Assessing the Impact of the Green Revolution, 1960–2000. *Science* 300: 758–762.

Eyzaguirre, Pablo and Evan Dennis. 2007. The Impacts of Collective Action and Property Rights on Plant Genetic Resources. *World Development* 35(9): 1489–1498.

FAO. 1996. *The State of the World's Plant Genetic Resources for Food and Agriculture.* Rome: United Nations.

FAO. 2010. *The Second Report on the State of the World's Plant Genetic Resources for Food and Agriculture.* Rome: United Nations.

Fowler, Cary and Pat Mooney. 1990. *Shattering: Food, politics, and the loss of genetic diversity.* Tucson: The University of Arizona Press.

Frankel, O.H. 1970. Preface. In *Genetic Resources in Plants—Their exploration and conservation*, ed. O.H. Frankel and E. Bennet, 1-4. Oxford: Blackwell Scientific Publications.

Frankel, O.H. 1988. Genetic Resources: Evolutionary and Social Responses. In *Seeds and Sovereignty: The use and control of plant genetic resources*, ed. Jack Kloppenburg, Jr., 19–46. Durham: Duke University Press.

Frankel, O.H. and E. Bennet. 1970. Genetic Resources. In *Genetic Resources in Plants—Their exploration and conservation*, ed. O.H. Frankel and E. Bennet, 7–17. Oxford: Blackwell Scientific Publications.

Fu, Yong-Bi, Gregory W. Peterson, Graham Scoles, Brian Rossnagel, Daniel J. Schoen, and Ken W. Richards. 2003. Allelic Diversity Changes in 96 Canadian Oat Cultivars Released from 1886 to 2001. *Crop Science* 43: 1989–1995.

Fu, Yong-Bi and Daryl J. Somers. 2009. Genome-Wide Reduction of Genetic Diversity in Wheat Breeding. *Crop Science* 49: 161–168.

Global Alliance for the Future of Food. 2016. *The Future of Food: Seeds of resilience, a compendium of perspectives on agricultural biodiversity from around the world.* https://futureoffood.org/report/the-future-of-food-seeds-of-resilience/ (accessed March 5, 2018).

Guarino, Luigi. 2012. "Brothers in farms." Agricultural and Biodiversity Weblog. https://agro.biodiver.se/2012/05/brothers-in-farms/ (accessed March 5, 2018).

Hammer, Karl, H. Knüpffer, L. Xhuveli and P. Perrino. 1996. Estimating Genetic Erosion in Landraces—Two case studies. *Genetic Resources and Crop Evolution* 43: 329–336.

Hammer, Karl and Gaetano Laghetti. 2005. Genetic Erosion—Examples from Italy. *Genetic Resources and Crop Evolution* 52: 629–634.

Hammer, Karl and Yifru Teklu. 2008. Plant Genetic Resources: Selected issues from genetic erosion to genetic engineering. *Journal of Agriculture and Rural Development in the Tropics and Subtropics* 109(1): 15–50.

Hardon, J.J. 1996. Conservation and Use of Agro-biodiversity. *Biodiversity Letters* 3: 92–96.

Harlan, H.V. and M.L. Martini. 1936. Problems and Results in Barley Breeding. In *United States Department of Agriculture Yearbook of Agriculture*, 303–346. Washington D.C.: United States Government offices Printing Office.

Harlan, Jack. 1975. Our Vanishing Crop Genetic Resources. *Science* 188(4188): 618–621.

Heald, Paul and Susannah Chapman. 2009. Crop Diversity Report Card for the Twentieth Century: Diversity Bust or Diversity Boom? http://ssrn.com/abstract=1462917

Heald, Paul and Susannah Chapman. 2009. Patents and Vegetable Crop Diversity (November 16). UGA Legal Studies Research Paper No. 09-017. http://ssrn.com/abstract=1507228

Heald, Paul and Susannah Chapman. 2010. Apple Diversity Report Card for the Twentieth Century: Patents and other sources of innovation in the market for apples. UGA Legal Studies Research Paper No. 10-01. http://ssrn.com/abstract=1543336

Heald, Paul and Susannah Chapman. 2012. Veggie Tales: Pernicious Myths about Patents, Innovation and Crop Diversity in the 20th Century. *University of Illinois Law Review* 4.

Hunn, Eugene. 1999. The Value of Subsistence for the Future of the World. In *Ethnoecology: Situated knowledge/located lives*, ed. Virginia Nazarea, 23–36. Tucson: The University of Arizona Press.

Imbruce, Valerie. 2007. Bringing Southeast Asia to the Southeast United States: New forms of alternative agriculture in Homestead, Florida. *Agriculture and Human Values* 24: 41–59.

Jarvis, Devra I., Anthony H.D. Brown, Pham Hung Cuong et al. 2008. A Global Perspective of the Richness and Evenness of Traditional Crop-Variety Diversity Maintained by Farming Communities. Proceedings of the National Academy of Sciences 105(14): 5326–5331.

Khlestkina, E.K., X.Q. Huang, F.J.-B. Quenum, S. Chebotar, M.S. Röder, and A. Börner. 2004. Genetic Diversity in Cultivated Plants—loss or stability? *Theoretical and Applied Genetics* 108: 1466–1472.

Love, Brian and Dean Spanner. 2007. Agrobiodiversity: Its value, measurement, and conservation in the context of sustainable agriculture. *Journal of Sustainable Agriculture* 31(2): 53–82.

Montenegro de Wit, Maywa. 2016. Are we losing diversity? Navigating ecological, political, and epistemic dimensions of agrobiodiversity conservation. *Agriculture and Human Values* 33(3): 625–640.

Mooney, Pat. 1980. *Seeds of the Earth: A private or public resource?* Ottawa: Food First.

Mooney, Pat. 1983. The Law of the Seed. *Development Dialogue* 1–2: 3–173.

Nabhan, Gary. 1989. *Enduring Seeds: Native American agriculture and wild plant conservation*. Tucson: The University of Arizona Press.

Nally, David and Stephen Taylor. 2015. The Politics of Self-Help: The Rockefeller Foundations, philanthropy and the 'long' Green Revolution. *Political Geography* 49: 51–63.

Nazarea, Virginia. 1998. *Cultural Memory and Biodiversity*. Tucson: The University of Arizona Press.

Nazarea, Virginia. 2005. *Heirloom Seeds and Their Keepers: Marginality and memory in the conservation of biological diversity*. Tucson: The University of Arizona Press.

Nazarea, Virginia. 2017. Landscapes of Loss and Remembrance in Agrobiodiversity Conservation. In *Routledge Handbook of Agricultural Biodiversity*, ed. Danny Hunter, Luigi Guarino, Charles Spillane, and Peter C. McKeown, 604–611. London: Routledge.

Nuijten, Edwin. 2005. Farmer Management of Gene Flow: The impact of gender and breeding system on genetic diversity and crop improvement in The Gambia. PhD Dissertation, Wageningen University.

Pistorius, Robin. 1997. *Scientists, Plants and Politics: A history of the plant genetic resources movement.* Rome: International Plant Genetic Resources Institute.

Ray, Janisse. 2012. *The Seed Underground: A growing revolution to save seed.* White River Junction, VT: Chelsea Green Publishing.

Seed. 2018. The Untold Story Discussion Guide: Where have all the purple tomatoes gone? https://drive.google.com/file/d/157RuVyhM1hPgtpFL0c6RyXrSjNRQNT4Y/view (accessed March 28, 2018).

Scarascia-Mugnozza, G.T. and P. Perrino. 2002. The History of *ex situ* Conservation and Use of Plant Genetic Resources. In *Managing Plant Genetic Diversity,* ed. J.M.M. Engels, V. Ramanatha Rao, A.H.D. Brown, and M.T. Jackson, 1–22. Oxon, UK: CABI Publishing.

Siebert, Charles. 2011. Food Ark. *National Geographic Magazine.* http://ngm.nationalgeographic.com/2011/07/food-ark/siebert-text (accessed February 2, 2018).

Shiva, Vandana. 1993. *Monocultures of the Mind: Perspectives on Biodiversity and Biotechnology.* London: Zed Books Ltd.

Shiva, Vandana. 1997. *Biopiracy: The plunder of nature and knowledge.* Boston: South End Press.

Shiva, Vandana. 2015. *The Vandana Shiva Reader.* Lexington: The University Press of Kentucky.

Teeken, Béla, Edwin Nuijten, Marina Padrão Temudo et al. 2012. Maintaining or Abandoning African Rice: Lessons for Understanding Processes of Seed Innovation. *Human Ecology* 40: 879–892.

Tribe, David. 2012. "Seed biodiversity in terms of numbers of commercial varieties of vegetables for purchase is largely unchanged over 100 years." GMO Pundit. http://gmopundit.blogspot.com.au/2012/06/seed-biodiverity-as-measured-on-in.html (accessed March 5, 2018).

van de Wouw, Mark, Theo van Hintum, Chris Kik, Rob van Treuren, Bert Visser. 2010. Genetic Diversity Trends in Twentieth Century Crop Cultivars: A meta analysis. *Theoretical and Applied Genetics* 120: 1241–1252.

Veteto, James. 2007. The History and Survival of Traditional Heirloom Vegetable Varieties in the southern Appalachian Mountains of western North Carolina. *Agriculture and Human Values* 25: 121–134.

Whealy, Kent. 2001. *Fruit, Berry, and Nut Inventory: An inventory of nursery catalogs listing all fruit, berry, and nut varieties available by mail order in the United States,* 3rd ed. Decorah, Iowa: Seed Savers Exchange.

Whealy, Kent. 2005. *Garden Seed Inventory: An inventory of seed catalog listing all non-hybrid vegetable seeds available in the United States and Canada,* 6th ed. Decorah, Iowa: Seed Savers Exchange.

Zimmerer, Karl. 1997. *Changing Fortunes: Biodiversity and peasant livelihood in the Peruvian Andes.* Berkeley: University of California Press.

Index

9781032338071